Farming and Famine

Africa and the Diaspora:
History, Politics, Culture

Thomas Spear, Neil Kodesh,
Tejumola Olaniyan, Michael G. Schatzberg,
and James H. Sweet

Series Editors

Farming and Famine

Landscape Vulnerability in Northeast Ethiopia, 1889–1991

Donald E. Crummey

Edited by James C. McCann

THE UNIVERSITY OF WISCONSIN PRESS

Publication of this book has been made possible, in part,
through support from the Anonymous Fund of the College of Letters and
Science at the University of Wisconsin–Madison.

The University of Wisconsin Press
1930 Monroe Street, 3rd Floor
Madison, Wisconsin 53711-2059
uwpress.wisc.edu

3 Henrietta Street, Covent Garden
London WC2E 8LU, United Kingdom
eurospanbookstore.com

Printed in the United States of America

This book may be available in a digital edition.

Library of Congress Cataloging-in-Publication Data

Names: Crummey, Donald, author. | McCann, James, 1950- editor.
Title: Farming and famine : landscape vulnerability in northeast Ethiopia, 1889–1991
/ Donald E. Crummey ; edited by James C. McCann.
Other titles: Africa and the diaspora.
Description: Madison, Wisconsin : The University of Wisconsin Press, [2018] |
Series: Africa and the diaspora : history, politics, culture
Identifiers: LCCN 2017051215 | ISBN 9780299316303 (cloth : alk. paper)
Subjects: LCSH: Agriculture—Environmental aspects—Ethiopia. | Agriculture—
Ethiopia—History—20th century. | Famines—Ethiopia—History. | Landscape
changes—Ethiopia.
Classification: LCC S589.76.E8 C78 2018 | DDC 338.10963—dc23
LC record available at https://lccn.loc.gov/2017051215

CONTENTS

ILLUSTRATIONS

Photographs taken by Armando Maugini on behalf of the Istituto Agronomico dell'Africa Italiana are used by permission of Agenzia Italiana per la Cooperazione allo Sviluppo.

Figures

Tables

EDITOR'S PREFACE

The basis of this book is the research and writing of Donald Crummey (1941–2013), a distinguished scholar of Ethiopia, who undertook and carried out his research with support from the MacArthur Foundation, the National Humanities Center, and the University of Illinois, and who unfortunately died before he was able to bring it to completion. This book is primarily his work, but it is also a collaboration to honor his memory, his work, and his colleagues and students, as well as his own voice.

The volume's content and tone reflect conversations I had as colleague and friend with the editors of the University of Wisconsin Press, especially with series editor Thomas Spear. Our goal here is to present in the core chapters Donald's literary and historian voice as well as the evidence he collected for this valuable body of research. The chapters include the work of an impressive set of scholars of Ethiopian history, ecology, geography, and natural science. His summary of that work is inclusive, even if at times it shows a tolerance for somewhat contradictory methods and interpretations of the progenitors of famine, landscape change, and the dynamics of local ecologies and agrarian livelihoods. The study areas include parts of Wallo and parts of contiguous landscapes of the Gondar region of Amhara province. The book focuses primarily on the core historical area of "Abyssinia" rather than the current national boundaries of Ethiopia. It recounts Christian Ethiopia and the Amhara people's historical experience but also ought to include the historical experience of peoples of Wallo, whose Muslim identities and full names express that they were part of Wallo's history. That unique mix of religion, ethnicity, and language is for many what it means to be from Wallo. Wallo's population also includes Oromo, Agaw, and Tigray peoples who do not appear explicitly, although they are part of the full story engaged by others. In focusing on these areas, the book benefits from the depth and breadth of Crummey's lifetime of work and the geographic and cultural foci of this study. It is a significant work of devotion and substance to Ethiopia's history.

As the book's editor and author of this preface and the introduction, I have sought to present Crummey's work and perspective on the project that he envisioned and assembled in nearly finished form. Though Don's precision in laying out evidence and argument is evident, at certain points tables were not yet complete, and at others he noted himself that a passage was not fully developed. I have at times omitted a paragraph, a table, or a small section.

The strength of this book is in the persuasion of Crummey's narrative. Its human qualities, in my view, are poignant, especially in the life histories that appear in the appendix and reveal the texture of selected lives lived on Wallo's and adjacent landscapes as described to the author and his team by the farmers—men and women—who offered testimony about their lives and memories of past livelihoods. If several of the voices are from contiguous areas not far away from Wallo, they may well reflect experience from contrastive agro-ecologies from the hot, dry periphery to the cool highlands. The selected cases demonstrate the variety of human experience from the region as a whole.

The introduction of the original book manuscript was not yet finalized. Don had wisely left the introduction and conclusion until the end of the writing. They were the unfinished part of the symphony. I tried to fashion the bookends in a way that represents Don's intentions and the contradictions that inevitably appear.

NOTE ON TRANSLITERATION

In consultation with the University of Wisconsin Press editors and in line with other publications on Ethiopia, this book's transliteration of place-names, historical actors, and historical concepts eliminates use of diacritical markings, except when authors included them in bibliographic references. In using Latin script, the goal is to help general readers reproduce pronunciation as effectively as possible. In the text, virtually all of the place and personal names are in the Amharic or Tigrinya languages and virtually none in Oromifa or other Ethiopian languages. Inevitably, there are inconsistencies. This system as used here, for instance, creates some problems with vowels, especially in the first and fifth orders, and in pharyngeal sounds from Tigrinya and Ge'ez. This volume renders vowels thus:

Transcription (Order in Ethiopic Script): Pronunciation
a (first order): as in "ever"
u (second order): as in "crude"
i (third order): as in "elite"
a (fourth order): as in "father"
e (fifth order): as in "bait"
e (sixth order): as in "fit"
o (seventh order): as in "vote"

In the transcription of consonants, *q* represents the ejective *k* and the palatized *ñ* (as in "canyon") is written with *ny*. Double consonants (as in "Wallo") indicate consonant gemination.

Personal names create the greatest discrepancies in transliteration, because many Ethiopians choose to spell their own names in print or in everyday usage in Latin script in ways that do not conform to a single system (e.g., Tadesse and Taddesse, Berhanu and Berhannu). Where an individual has already used a certain spelling of their name in Latin script, we used that. In certain public spellings in Latin script (e.g., Haile Selassie versus Hayla Sellase), we accept those conventions for those individuals as they have expressed their spelling choice in Latin script. For place-names we followed the above transcription unless there is a more or less common Latin spelling (Dessie rather than Däsē or Däsé).

Ethiopian names consist of a given (first) name and a father's name. In the text, the name used after the first full citation is the given name. In the bibliography Ethiopian names are alphabetized according to the first name.

Farming and Famine

INTRODUCTION

James C. McCann

FOR THE STUDY of landscape change and its historical setting, Ethiopia is both a treasure and a conundrum—a setting for hope and episodic tragedies. In the "long" twentieth century that is the focus of this book (1889–1997), Ethiopia's people have endured drought, disease, and consequent hunger that affected the "salubrious" highlands and adjacent lowland periphery. In the eastern periphery, historically settled and more recent populations migrated onto marginal lands and witnessed movements of political resistance (banditry) in opposition to the politically dominant highland plow-based economy. Those lands absorbed, sometimes by force of state policy, new populations that settled on landscapes vulnerable to drought and occupied the margins of the national political economy. These are long-standing patterns.

Historians and scholars of Ethiopia have long struggled to understand the "Ethiopian paradox": how could Africa's most productive food production system (Ethiopia's ox-plow, annual cereal agriculture), which sustained an extraordinary imperial culture over two millennia, also be home to periodic tremendous famine and rural poverty? In modern times Ethiopia has suffered three grievous famines, two of which—in 1973–74 and in 1984–85—captured world attention. These famines affected the lives of millions of farmers and pastoralists. In fact, in the late twentieth century Ethiopia has surpassed earlier icons of famine: China, India, Armenia, and Biafra. Ironically, Ethiopia's highland culture also generated and eventually exported the iconic cuisine served in Ethiopian restaurants throughout the developed world and in large African cities.

Common resolutions of this paradox point to the landscape, where it is said to be evident that the population increase has driven Ethiopia's farmers to mine their environment, plowing ever steeper slopes and cutting down trees for construction and fuel. These have combined to create soil erosion, which is undermining farmers' livelihoods. This perception is folded into globalizing narratives of deforestation and desertification, which thereby become processes held to be central to understanding famine vulnerability in Ethiopia.

Donald Crummey challenges such narratives. In the face of increasing environmental stress, he argues, Ethiopian farmers have innovated and adapted. In the process they have developed effective strategies for managing their environment—strategies often ignored by conservation projects. Crummey's final paragraph in this book makes that observation clearly, and with passion: "A more perplexing question is how farmer initiatives in general . . . have gone so unnoticed; and how Malthusian rhetoric has acquired such a purchase on explanations of landscape change. Governmental and elite commitments to global narratives of deforestation and desertification have obscured what is happening in open daylight. However, if one fails to ask the question, one cannot expect to get the answer."

So, Crummey asks, how do we assemble a tragic theater of famine and landscape change, a human story that evolves over time and then pounces at key moments such as the Great Famine (Kefu Qan) and subsequent crises in the late 1920s, 1958, 1972–74, 1984–86, and even more recently? The deeper question about these events is: how could Ethiopia's rural people—farmers, herders, part-time traders, and local political actors—dedicate themselves to maintaining technologies, local knowledges, and social lives on a rural landscape that mixed verdant highlands, dry lowlands, and dramatic seasonal shifts in moisture? In the late 1980s and early 1990s, when I asked farmers in my work for Oxfam and Redd Barna (Norwegian Save the Children) why they planted certain crops on particular fields, those farmers often responded fatalistically, "Maretu endi mert!" (as the land chooses). The sophisticated agricultural ecology evident in Ethiopia's landscapes was, in their view, a result of the guidance—and serendipity—of nature's own eccentricities.

Crummey addresses these questions in terms of a historical study of famine on the Christian highlands and a detailed vision of the Wallo region—one part of which anthropologist Donald Donham calls "old Abyssinia and the new Ethiopia empire." The first three chapters recount the history of the Great Famine of 1888–92 and the famines of the mid-1970s and mid-1980s. Together these chapters describe the local theater of famine in Wallo and offer a general consideration of environment, famine, and local responses to it in the Wallo region and some nearby highland areas. As Crummey explains, international agencies and Ethiopian officials viewed the famines of 1973 and 1984 through the lenses of deforestation and desertification. These processes brought a predetermined set of interventions, most prominently large-scale tree planting and anti-erosion measures, which bore heavily on farmers, many of whom found themselves transplanted to different landscapes through resettlement of farming populations.

The remaining chapters offer Crummey's Boserupian counterpoint to what he characterizes as the dominant Malthusian thesis on the driving forces for famine and poverty in Ethiopian central highland famines of the late nineteenth and twentieth centuries. His central, recurrent thesis is about a resilient rural population. Focusing on Wallo, he presents the evidence of the historic landscape as recorded in a set of photographs taken in 1937. These photographs are assessed in light of rephotography—the production of a fresh set of photographs in 1997, which replicate the perspective of the originals as closely as possible. Laying out this powerful evidence, Crummey offers a rejoinder to degradation narratives that have dominated many accounts of Ethiopia's environmental past. To uncover the dynamics of change, he presents life histories from some of the nearly ninety interviews he conducted with men and women who lived in and through these landscapes.

Wallo is only a portion of what we now include under the complex ecologies of Ethiopia. It is a political region with boundaries marked by local identities that over time combined Christianity and Islam, highland ecologies and eastern escarpments, leading to dry scrubland vegetation more suited to livestock rearing than to intensive agriculture. Crummey's career fascination with the northern highlands, including parts of Wallo, offers a lens on a Christian kingdom that had fallen into the periphery of the Addis Ababa political economy in the twentieth century. In this sense, Wallo is a particularly interesting region from which to view the culture and ecology of landscape change, national politics, and vulnerability to famine. Activist geographer Mesfin Wolde Mariam, a native of southern "Wello," exclaims in his book (cited by Crummey) that "love reigns supreme in Wello," meaning that the area's history includes cultural and political interactions between Islam and Christianity. Both groups could hold land there. They even have a distinctive practice of giving their children Muslim and Christian names, evident across generations among neighbors and often within single family lines. Moreover, the people that we might now classify as Wallo encompass a collection of identities changed over the long twentieth century, crossing political boundaries that sometimes included historically distinct areas like Wag and Lasta, Were Ilu, Wadla-Delanta, Dessie Zuriya, Tehundare, and so on. Thus, places and identities slipped in and out of the formal political boundaries of Wallo along with the political winds of each era. This book covers some (but not all) of these. Nonetheless, as Mesfin points out, Wallo's history and geography, economic networks, and cultural identities include the river valleys that drain the highlands into the dry lowlands of the Mille and Awash valleys, where pastoral cultures absorbed highland farmers whose arrival after World War II reshaped who was vulnerable to famine. These were dynamic local economies (and population movements). Ironically, in the post–World War II era much of the population who moved to famine-prone areas did so by force or perhaps in the hope that their farms might escape the inevitable ravages of drought.

Beyond the in-depth presentation of historical evidence on the northern highlands around Wallo, this study compares landscape photos taken by the Italian Armando Maugini in 1937 with similar ones taken by Crummey's team in 1997, allowing us to compare changes over the intervening decades. The insights on vegetative cover, human settlement, and agricultural fields are remarkable, adding weight to earlier comparative fieldwork done by Pauline Boerma in Eritrea, which compared photographs of 1930s landscapes to residents' memories of landscape vegetation. Those photographs, she found, showed that local folk memories were inaccurate and residents had adopted degradation perspectives from government and nongovernmental organization policy literature: they told interviewers not what they themselves remembered but what they had learned was the accepted story.[1] In Crummey's book, the life narratives of Wallo and south Gondar farmers—men and women—provide a valuable complement to Boerma's field studies. By lining up photographed sites to compare landscape cover, Crummey's work offers precision and depth in assessing how forest cover changed (or did not) over the course of half a century in part of the central highlands.

In truth, Crummey's erudition, drawn from historical and detailed compilation of more recent studies, is a study of uplands (*dagga* and *wayna dagga*) and the ethnographic variation of Wallo, though sometimes missing its peripheries at the edge of the escarpment extending from Tigray south to northern Shawa. In those lower zones, periodic drought was more common, some crops (i.e., sorghum) were more commonly cultivated, and more diversity of livelihood/language/religion marked the human landscape. In Wallo, unlike Gondar and Gojjam, Muslims could and did cultivate rural land, especially along and below the escarpment. Many of these groups migrated south into areas of the southwest during the coercive resettlement program of the late 1980s. Human movement, crop strategies, and interventionist government policies made for a dynamic human ecology that engaged more just than the highland zones.

Forest or woody cover was a part of this mix, and changes there were more visible than those of annual crops. Crummey's study addresses this fact, though forest cover has different meaning and economic signals. Hard woods like juniper or podocarpus and acacia wooded landscapes had very different uses than the increasingly common eucalyptus, as Crummey notes. Most new tree plantings across the highland and midrange areas consisted primarily of eucalyptus plantations. The latter is a poor source of fuel wood or charcoal compared with hardwoods like acacias, juniper, ficus, and podocarpus but has great value for building or fencing.[2] It makes a charcoal of last resort. We might have questions about the ecology of the forest cover evident in the photographs and in the life stories as told by the interviewed farmers. These contrasting images and perceptions of Ethiopia's history are also part of the national historical heritage and part of a continuing dialogue.

Overall, Ethiopia's landscape history is a story of human adaptation via local and global migration, persistence of an economic culture (plow agriculture), and the complexity of moving elements worthy of quantum mechanics theory. One could also say that the study of Ethiopia's landscape history raises more questions than it answers. Let us recognize and appreciate that, with this book as a true measure of Don Crummey's curiosity, dedication, and intellectual leadership.

1

Famine in Ethiopia

The Great Famine of 1888–1892 Revisited

THE KEFU QAN, the Great Famine of 1888–92, is a good starting point for reflecting on famine in Ethiopia and the complex ways social, economic, political, and environmental factors interacted to bring it about. It is the earliest Ethiopian famine for which we have rich sources. The "Evil Days" or "Bad Days" are well remembered throughout northern Ethiopia. Seventy-year-old informants in Armachachaho district in 1997, who had no memory of the national famine of 1984, recalled hearing about the Great Famine of 1888–92 from their parents.[1] For all its importance, the Kefu Qan has not yet received the attention it warrants.[2] Its relevance to contextualizing famine in the twentieth century and the availability of new material notwithstanding, many facets of the Kefu Qan remain beyond our ken.

The Kefu Qan: A Chronology

In the early 1880s, Ethiopia seemed to be in an unusually strong position in the face of rising European imperialism. The nation had entered the nineteenth century bitterly divided, but Emperor Yohannes IV (r. 1871–89), building on the drive of his predecessor Tewodros II (r. 1855–68), had established his sovereignty over much of the historic highlands, notably over the prominent vassals, Takla Haymanot of Gojjam and Menilek of Shawa, and successfully resisted two attempts by the Egyptians in 1875 and 1876 to invade.[3] However, his success proved short-lived. His defeat of the Egyptians contributed to their rapid decline, and Yohannes soon found himself faced with not one but three energetic challengers, who had brought him close to the end of his tether by 1888. To the west he faced an aggressive foe in the Sudanese Mahdist state; to the north he was under pressure from expansionist Italians, who with the connivance of the British in 1885 had occupied the island port of Massawa; to the south was Menilek II, who was in league with Negus Takla Haymanot and in open defiance.

In March 1888, Yohannes besieged a small Italian garrison at Sahai, inland from Massawa. Then, distracted by Menilek's rebellion and by Mahdist incursions into the central highlands, he broke off the siege in April and marched south, carrying cattle, spoils of his war with the Italians, in his train. The Italians had imported these cattle, probably from India, and they brought rinderpest, a disease "enzootic in the cattle of the steppes of eastern Asia," hitherto unknown in tropical Africa.[4] It had a subcontinental impact.[5] As Yohannes marched south into Begemder and Gojjam, which he reached toward the end of the rains, rinderpest followed him. Its impact on Ethiopia's population was devastating. Four years later, the full impact of the famine clear, the Italian explorer and medical doctor Traversi wrote, "I am not exaggerating in saying that one-third of the population has died and if things go on like this, probably many more, unfortunately, will die."[6] In virgin rinderpest epidemics, the death rate among cattle-like ungulates, domestic and wild, can range up to 95 percent.[7] Although some transhumant pockets of pastoralists managed to avoid infestation, many other societies were devastated. For Ethiopia's farmers, it meant the loss of a crucial factor of production—plow oxen—and this helped precipitate famine. For the Ethiopian scholar Alaqa Lamma, there was no doubt about the cause of the famine. It "was caused by the rinderpest which decimated cattle. With what to plow? The land was left fallow. The famine immediately followed the cattle disease."[8] Traversi agreed: "The first and only cause . . . [of the famine] . . . was the death of the animals."[9] Although rinderpest undoubtedly played a very large role in creating famine, the sources, as we will see, put other factors into play—drought, warfare, other epidemic diseases, and pest infestations—and the interplay among these factors is difficult to tease out.

The general course of the famine, which persisted from 1888 through to 1892, is reasonably well known, though not without its puzzles and uncertainties. There are both European and Ethiopian sources for reconstructing its course. These sources are spotty in their coverage through time and over space. European documents are particularly useful for the comparative precision of their dating and location, whereas Ethiopian documents are indispensable for dating and providing a local perspective. There are accounts of Ethiopia from Europeans representing at least five different foreign nationalities. Italians are most numerous—the period coincides with the declaration of their colony of Eritrea in 1890 and their attempts to curry favor in Addis Ababa.[10] Swiss entrepreneur Alfred Ilg enjoyed the confidence of Emperor Menilek and was an active correspondent.[11] French Catholic missionaries were firmly established in northern Ethiopia, while German-speaking Protestants kept in touch with Ethiopian converts around Gondar.[12] Finally, British consular officials in Aden and Suakin regularly reported on events on the African side of the Red Sea.[13] Despite this diversity, reporting concentrates around three locations: Eritrea and the port of Massawa; Shawa with the court of Menilek at Entoto and Addis Ababa and the Italian Geographical Society's station at Lat Marafiya; and Harar and the eastern highlands.[14] Among the Ethiopian sources two authors are preeminent—Alaqa Gabra Sellasie, Menilek's official chronicler; and Asma Giyorgis, a Catholic convert and author of "A History of the Galla"—although there are other chronicles, a considerable correspondence from Ethiopian authors, and memory.[15]

Though it is unlikely that full-fledged famine had yet broken out, there are reports of sometimes desperate hunger from 1888. For example, the missionary Crouzet wrote in January 1889 of famine, which had been going on for over a year.[16] Caulk, together with other sources, refers to "penury" and "famine" within Yohannes's army as early as March 1888, and his withdrawal from Sahai, although this might as easily have indicated contingent shortages as it did the onset of the Kefu Qan.[17] Count Antonelli, an Italian diplomat assigned to Menilek's court, had informants within Menilek's army, which at Yohannes's behest was stationed outside Gondar to protect the central provinces against Mahdist incursions. Sometime between January and March, Antonelli's informant wrote, "Because our entire army is suffering from hunger, the soldiers have been sent to *caleb* [*sic*] (that is to be provided on a daily basis with food from the houses of the peasants)." These same troops were dispersed widely, from west of Lake Tana northward to the Takaze River.[18] That same informant wrote on May 17 that Yohannes's army, too, was "famished and crying out for mercy. All the soldiers said they would rather die fighting, than to die from hunger."[19] Following the rains, sometime in October 1888 Yohannes wrote to Menilek concerning the latter's daughter, who had been married to Yohannes's recently deceased son: "I had intended to send your daughter . . . on October 12, but, since I now lack the funds and do not want to send her with empty hands, it will be in

November." Rosenfeld attributes this lack of funds to the famine.[20] However, no other source confirms famine within Yohannes's or Menilek's army following the rains of 1888.[21]

The outbreak of the Kefu Qan is generally traced to October 1888 on the evidence of an anonymous letter quoted in *Les Missions Catholiques* by Bettembourg, the procurer of the Lazarist missionary order in Paris. Bettembourg's letter is dated Paris, November 16, 1888, so the events to which it refers probably occurred not long before, perhaps toward the end of October, when farmers would be looking forward to their primary annual harvest. The anonymous letter laments the condition of "our poor Catholics," communities on the Eritrean plateau: "To the continual fires and pillaging which had practically ruined them, has been added an epidemic which has taken a large part of their herds. To add to the misfortune, the harvests failed because of the lack of rain and famine began to take hold in the districts of our Christians."[22] This description is echoed by an account of the origins of the famine by Gustav Arén based on records of Swedish Evangelical missionaries, who were present in Eritrea at this time: "It had begun in 1888 with a long period of drought in the north, where provisioning units from the imperial army had denuded the area of all supplies. The drought led to harvest failure followed by an outbreak of rinderpest, which carried off the greater number of the cattle. The killing off of almost all the plough oxen and the shortage of grain for sowing drastically reduced agriculture. The growing crops were completely destroyed by swarms of locusts and masses of caterpillars."[23] Both missionary sources indicate that famine hit Eritrea first. It was well established by January 1889, when the Catholic missionary Crouzet observed, "The inhabitants of Keren and the interior have nothing to eat and nothing is growing."[24] For over a year, he wrote, famine had been accompanied by "all manner of scourges." First the epizootic had killed the animals, then in some places, "all the harvests have been burned in the field by the sun or devoured by locusts and caterpillars."[25] Three months later, in surveying the prospects for the occupation of Eritrea, the commanding officer of the Massawa garrison wrote to the Italian Ministry of War, noting that "the great famine would make our permanent occupation of the [Eritrean] plateau difficult."[26] That February, Italian diplomat Antonelli, assigned to Menilek's court, had reported from Addis Ababa that "Tigre, Gondar, Begemder, and Gojam are presently in the most squalid misery" because of the lack of grain "and because of the death of cattle from disease." His report went on to note that although famine had not yet hit Shawa, an "epidemic disease" had destroyed all of its cattle. "Entire herds of 500 and 1000 head have been destroyed in one or two days."[27] He also remarked on "epidemic illness" within Yohannes's army, then menacing Shawa. He anticipated that the loss of oxen at this critical juncture of the agricultural year would compromise plowing and produce famine.[28] His account accords well with the memory of Alaqa Lamma, who recalled that although the cattle had died, as of May 1889, famine had not yet struck northern Shawa.[29]

Apparently this was no longer the case as the rains set in. Vincenzo Ragazzi, who had served as director of the Italian Geographical Society's station at Lat Marafiya in northern Shawa, reported that a "deadly" typhus epidemic "which has almost completely destroyed the cattle [illegible] now, in association with a very grave form of dysentery, menaces destroying the population." He went on to report that on July 27, 1889, Menilek issued a proclamation, acknowledging that "the epidemic has turned to people and has begun to destroy them."[30] At the same time, rinderpest had reached the eastern highlands. Writing to Ilg from Harar, French poet and gun-runner Arthur Rimbaud described conditions similar to those elsewhere in Ethiopia: "The epizootic has destroyed everything here, the coffee harvest is zero, the sorghum harvest mediocre."[31]

In the midst of this unfolding disaster, Ethiopian politics were roiled by the death of Emperor Yohannes IV on March 10, in battle with the Mahdists.[32] About the famine, the Shawan chronicler Gabra Sellasie wrote: "In that time, the year of Matthew [Sept. 1888–Sept. 1889], when *keremt* [the long rains, from late June to mid-September] arrived, all the cattle had been exterminated by disease. . . . Also, the following year, the year of Mark [Sept. 1889–Sept. 1890], a general famine broke out."[33] Entoto, the capital of the country's new ruler, Menilek, was besieged by hungry throngs. One week before Menilek's coronation in nearby Addis Ababa on November 3, 1889, Alfred Ilg, a Swiss entrepreneur and confidant of Menilek, wrote that "people are at a loss for food."[34] Following the ceremony, Menilek marched north to Tigray to subdue Yohannes's would-be successor, Mangasha. Given the wretched state of Tigray, he abandoned plans for a second coronation at Aksum. So virulent was the cattle epidemic, somewhere in the course of this year it reached Kaffa in the southwest, where, as elsewhere, its advent was followed by the deaths of many people.[35]

The year 1890 saw famine worsen. In January Crouzet again wrote from Massawa noting that the famine had spread from Bogos to Akala Guzay in eastern Eritrea.[36] In March another Catholic missionary, Coulbeaux, wrote from Bogos of general public misery and "walking skeletons."[37] That same month, his hungry and diseased army refusing to continue the campaign, Menilek withdrew from Tigray. In late April, he reached Dessie, where Traversi, summoned by the emperor because of his medical skill, reported to the Italian envoy Augusto Salimbeni, "the army [is] decimated by dysentery," accompanied by hunger.[38] As for Lat Marafiya, Traversi's place of residence, the station itself was well enough off, but the country in general was marked by famine. The roads leading to the houses of nobles and to Addis Ababa were lined with the dead and dying.[39] In April a correspondent wrote to Ilg and confirmed that famine existed in Tigray and had driven people from Adwa to Asmara in search of food.[40] Meanwhile, Salimbeni was on his way inland from Massawa to join Menilek passing through country already familiar from a previous trip. Early in May, still in Eritrea, he found throngs of emaciated people making their way to Massawa in search of food. The countryside was extremely desolate, without signs of cultivation.[41] Between Adwa and Maqale, he reported seeing "only one pair of oxen" yoked to a plow. "Of the fat and numerous herds of seven years ago, not a trace."[42]

Returning to Addis Ababa in June, Menilek repeatedly deferred expeditions to Tigray.[43] That same month, Ras Mangasha, de facto ruler of Tigray, wrote to the Ethiopian community in Jerusalem, apologizing that he could not help them with money, "as our country is stricken by famine. It is the period of tribulation."[44] In August, conversation in Addis Ababa recognized the importance of the cattle epidemic and expected intensified famine.[45] A small Protestant community near Gondar lost 177 of its members between May and October 1890, a reflection of the "indescribable" famine that raged through Tigray and Amhara.[46] In October 1890, Traversi claimed that famine had reduced people to cannibalism. That same month the "influenza" epidemic, which had killed 20,000 of Menilek's troops returning from Tigray, reached "virulent proportions" in Addis Ababa.[47] A month later, writing from Ankobar, not far from Lat Marafeya, Ilg reported to his correspondent that famine "has claimed many lives."[48]

The following year, 1891, was no better. The chronicler Asma Giyorgis noted that in the year of Luke, which ran from September/October 1890 to August/September 1891, there occurred great famine, which killed many people in Shawa and Wallo.[49] At the beginning of February 1891, Rimbaud wrote to Ilg from Harar, "In a few months we will have a terrible famine.

The sorghum harvest has failed."[50] Later that month, Salimbeni, returning to Italy via Harar, reported from the Deder highlands, "Desolation reigns on all sides." Closer to Harar he came on desperate famine refugees from Shawa.[51] In April, Ferdinando Martini, parliamentary deputy and member of an official commission of inquiry, wrote to his daughter from Massawa of people dying of hunger only a half kilometer from the town—"A spectacle which I will never forget"—and went on to describe the countryside being in the grip of an aridity without comparison.[52] Ilg, en route to Europe, traveled in Eritrea in spring 1891 and witnessed the continuing effects of famine.[53] Similar conditions apparently existed in Harar in June, according to information obtained in Aden, which noted that "severe" famine prevailed there.[54] In October, Traversi wrote again from Lat Marafeya: "The country is in a terrible moment. Famine, from day to day more acute."[55] Typhus, he added, also contributed to the fatalities. That same month, Menilek again postponed his planned second expedition to Tigray. The condition of Harar "was still very dreadful owing to the famine."[56] Aden reported to London that Italian sources told of a meeting in December 1891 between the governor of Eritrea, Gen. Gandolfi, and "the Chiefs of the Tigrian provinces," at which "the women and children accompanying the troops of Mangasha, . . . appeared haggard, emaciated and in a wretched condition."[57] Caulk notes that during this year, the lack of food forced some military leaders to break up their units, sending the soldiers in search of food; others led their troops to raid the Arsi, southeast of Addis Ababa, who apparently had escaped the worst effects of the rinderpest and epidemic human diseases.[58] In Harär, by the end of 1891, there was still a dense population and satisfactory harvests in the immediate surroundings, but "everywhere else, in the once flourishing eastern highlands, half or more of the population had died. Large tracts of land lay idle."[59]

In 1891 we find fresh indications of locusts, an affliction to which Crouzet had referred as early as January 1889. In the account of his 1891 trip, Ilg also mentions caterpillars (*Raupen*) and locusts, which both destroyed crops. The locust swarms were not confined to Eritrea. In September 1891, an Ethiopian correspondent wrote to Ilg from Entoto that Bulga, a district to the northeast of Addis Ababa, "has been eaten by locusts."[60]

Famine persisted well into 1892, along with reports of cholera. In a letter dated January 1892, Traversi reported that famine had spread to the Oromo lands to the west and southwest of Addis Ababa and to Harar and other parts of the eastern highlands. "Harar, then," wrote Traversi, "if one permits me the phrase, is today the emporium of all these miseries."[61] Ilg's correspondents in Shawa wrote mostly of the effects of epidemic disease, *koléra* in one letter, dated April 25, inducing the court to move from the Addis Ababa area to Ankobar to escape its effects.[62] In July the situation in Eritrea seemed as bad as ever: "The poor people, having nothing left, have been forced to go into exile. They have been condemned to die from hunger or to become the prey of wild beasts. People, in order not to die of hunger, sell their goats, their cows, their mules, their lands and everything which they own."[63] German explorer Georg Schweinfurth reported in the first half of 1892 that there was smallpox in Eritrea, a note that is repeated by Crouzet in a letter from Akala Guzay, dated July that same year.[64] Schweinfurth also mentions locusts continuing to be a problem in Eritrea, as do other sources.[65] Things were no better in Bagemder, where sometime during the main rains of 1892 (perhaps in August), Aragawi, a correspondent of Martin Flad, informed him, "Our famine and calamity are indescribable," went on to repeat the stories of cannibalism, and noted that an epidemic of cholera had affected his community during "the last rainy season."[66]

The harvest of late 1892 must have been adequate as the references to contemporary famine taper off by the end of that year, although famine remained a subject for retrospective comment for some years to come and lodged itself enduringly in popular consciousness. Already in April 1892, striking a different note from the one sounded from Aden, the British consul in Suakin reported to the foreign office that Ethiopian armies were again in the field, because "they find food."[67] That same month, Aden reported that Ras Makonnen, governor of Harar, had returned from an expedition to the Ogaden with a large herd of plundered cattle, many of which were sent on to Ankobar as famine relief.[68] Adjustments to production, such as the use of horses, mules, and donkeys as draft animals, must have proved adequate for the reduced demand from a diminished population. Cattle, like those seized by Makonnen from the Ogaden, also replenished oxen for the farm economy. However, in September, although it did appear that the harvest was promising and cholera was now found only in Ankobar, famine still prevailed "everywhere."[69] A month later, yet another Italian, traveler and businessman Luigi Capucci, could write from Entoto that the country was still "grievously famished."[70] Only in December could he indicate that events had taken a decisive turn for the better and that "no one is dying of famine any more and that a certain sense of well-being is making itself known throughout the country."[71] The other afflictions seem to have abated, although cholera and drought may have carried on in Eritrea into 1893.[72] Nonetheless, it was not until 1895 that Menilek felt the supply of food to have replenished itself to the point where he could undertake the northern campaign, which deepening famine had postponed.

The geographical reach of the Kefu Qan was unmatched by later famines. Contemporary observers placed it in Eritrea, Shawa, and Harar as well as Tigray, Bagemder, Gojjam, the Oromo lands to the west and southwest of Addis Ababa and to the south and southeast of Harar, and Kaffa, further to the southwest. Asma Giyorgis claimed, "The extermination of the cattle occurred everywhere from Masawa to Kafa, and from the 'Arab to Harar."[73] This observation is confirmed in some detail by European travelers who visited the Somali and Oromo territories, as well as southwestern Ethiopia in the immediate aftermath of the rinderpest.[74] Moreover, the rinderpest spread westward across the continent, reaching the Atlantic Ocean in 1892, and southward, reaching the Cape of Good Hope in 1897.[75] In many places, in Eastern, Southern, and Sahelian Africa, famine followed in its wake, its effects often intensified by other cattle diseases, smallpox, and locusts, plagues that also accompanied it in the Ethiopian highlands.[76] Given the role Italy played in introducing rinderpest to the continent, it is not surprising that elsewhere in Africa, it appeared in close association with the new imperialism of the 1890s. It undermined the capacity of some societies, like the Maasai, to resist the advance of European rule. Elsewhere it brought hardship, which in the case of Zimbabwe (Southern Rhodesia) helped provoke the Shona-Ndebele uprising of 1896–97, which contributed to ending the rule of the British South Africa Company.

Understanding the Kefu Qan

The chronology of the Kefu Qan gives some idea of its scale. But a chronology alone leaves many questions unanswered. How did environmental disturbance affect Ethiopia's agrarian economy? What role did drought play? How did rinderpest contribute to famine? What additional pests were at work? What epidemic human diseases were involved? How did they interact with drought and famine? What role did Ethiopia's ruling classes play in creating the conditions for and subsequently responding to this disaster? What effect did famine have on Ethiopian society?

Our sources for the outbreak of the famine are agreed that it broke out in Eritrea as a result of several factors: pillaging by marauding armies;

the loss of cattle to epidemic disease; a lack of rain, which brought harvest failure; and locusts and caterpillars, which destroyed any crops that may have been planted. Revisiting these factors one by one within the framework of the agricultural year should deepen our understanding of the dynamics of the Kefu Qan.[77]

The Ethiopian climate is typically subtropical in its seasons of wet and dry. The principal rains, which begin in June, bring Atlantic moisture with the great air masses, which are drawn northward over the continent by the movement of the Intertropical Convergence Zone (ITCZ). The location of the ITCZ in turn is determined by the how the angle of the Earth's axis of rotation to the plane of its rotation (currently about 23.5 degrees) steadily alters the angle at which the rays of the sun hit the surface of the Earth in the course of the annual orbit around the sun. The forces driving the monsoon patterns of the Indian Ocean and bringing spring and summer to the Northern Hemisphere also cause the northward movement of the ITCZ. In June the Atlantic air masses rise when they encounter the Ethiopian highlands and, in doing so, lose moisture. These rains bring the *keremt*, or rainy season, which lasts through to September, when the equinox approaches and the ITCZ starts its southward movement; the rains feed the major harvest. Throughout the rainy season, precipitation is heaviest in the southwest and declines steadily to the northeast.

Ethiopian agriculture is the product of 3,000 to 4,000 years of experiment and development, finely tuned to an extraordinary array of distinct microenvironments, the products of differences in altitude, soils, exposure to wind and sun, and availability and timing of moisture. Ethiopian farmers grow a wide variety of cereals and pulses: wheat, barley, peas, beans, chickpeas, lentils, of Middle Eastern origin; sorghum and finger millet of African origin; and *tef*, a cereal of local origin. Oil plants are also important, most of them of Middle East origin—safflower, sunflower, sesame—but *nug* is local. These crops provide the elements for a developed cuisine, which also draws on herbs, spices, and animal and plant protein. Based in the highlands of the eastern and southeastern basin of the Nile, the historic Ethiopian state held sway over a great variety of peoples, with distinct agricultural practices, but it was embedded in an agrarian economy based on cultivating cereals and pulses with the ox-drawn plow.[78] Stalks of wheat appeared on the coins issued by the Aksumite rulers of the early Christian era.

The lives of the rulers and subjects of historic Ethiopia played themselves out within two annual cycles determined by the liturgical calendar of the Ethiopian Orthodox Church and the agricultural calendar, shaped by climate. The Ethiopian agricultural year starts in late April or May, in anticipation of the onset of *keremt*, with plowing and planting increasing in intensity into July, in anticipation of a principal harvest in November and December. Farmers mark the agricultural year in terms of the liturgical calendar with its cycles of monthly and annual festivals and fasts. Planting varies according to crop and agro-ecological zone. Highland Ethiopians perceive three distinct agro-ecological zones—*dagga*, highland; *wayna dagga*, middle (lit. "grape") highland; and *qolla* lowland.[79] Although these do have objective correlates—at 3,000 meters above sea level you are definitely in *dagga*, at 2,000 meters is *wayna dagga*, and at 1,000 meters *qolla*—they are relative terms, and each carries cultural overtones. Some crops do better in one zone than in another—sorghum is overwhelmingly the predominant cereal in the *qolla*. At the highest elevations of cultivation, barley is the only cereal that will grow.[80] The widest variety of crops grows in the *wayna dagga*. Within each agro-ecological zone planting also varies according to soil conditions and the local microenvironment. Whereas the principal harvest arrives in November and December, some parts of the highlands enjoy sufficient rain around February and March to provide for a second (*balg*) harvest. It is no accident that the earliest indication of famine came in October 1888 in anticipation of the failure of the main harvest of that year, nor that the clearest indication of its end came in December 1892, following that year's main harvest.

Rinderpest

Epidemic cattle disease played a major role in the famine and is testified to by numerous observers at different locations throughout the period. Rinderpest is a viral disease that attacks ruminants, wild and domestic, particularly water buffalo and cattle, as well as sheep and goats (less severely).[81] There seems to have been no uncertainty about its identification. The sources use a variety of terms in referring to it (see table 1.1). The word "rinderpest" is German in origin. The customary English equivalent is "cattle plague," which, despite its generic appearance, does indeed refer to a distinct disease.[82] Europeans were familiar with the disease, an outbreak in the late eighteenth century having led the French government to establish its first veterinary college.[83] Cattle plague entered Great Britain in 1851 from the Baltic states and led to the creation of a Cattle Plague Department of the Privy Council.[84] In a virgin situation, animal mortality may

TABLE 1.1. Terms used to refer to epidemic cattle disease

Language	*Term*	*Source*
Amharic	bashata	Bairu Tafla, *Asma Giyorgis and His Work*, 804
	ma'at	Gabra Sellasie, 149
French	épidémie	Bettembourg, 557
	épizootie	Crouzet (1889), 62
	grand fléau	Abba Fezzah, 303
	peste bovine	Wurtz, "Hygiène Publique et Privée en Abyssinie," 492
Italian	epidemia tifica	Ragazzi, *Boll.* 1889, 964
	epizoozia	Traversi, "Lat Marafeya," 228
German	Viehseuche	C. Keller, *Alfred Ilg*, 89
English	disease	Stace (1893), 9
	epidemic of epizootic	Skinner, *Abyssinia of Today*, 195
	rinderpest	Wylde, *Modern Abyssinia*, 339

reach 90 percent and may drop to 20 percent in recurrent epidemics. Although there is some disagreement among the authorities on the issue, it is most likely that Ethiopia had had no prior exposure.[85] In any event, the epizootic behaved exactly like a virgin episode, and the 90 percent figure is consonant with contemporary impressions of cattle mortality.

The virus spread with great rapidity. Yohannes had left Sahai in April 1888.[86] In May the cattle of Hamasen on the Eritrean highland were dead, as were the cattle of Tigray by June. There is a hiatus in reports of the epidemic until January 1889, when Alaqa Lamma recalls going to sleep early that month and waking the next morning to find that all the cattle of Gojjam had died.[87] By February 1889, as we have seen, the cattle of Shawa had been decimated.[88] The great rinderpest epidemic destroyed a major resource on which Ethiopians of all ranks were dependent. The Amharic word for wealth, *kabt,* is also the word for cattle. Cattle were important to Ethiopia's rulers, who kept large herds. Cattle were the most important form of booty in their frequent expeditions against neighboring peoples. The ostentatious consumption of beef was central to the banquets held by Ethiopia's rulers, although at the other end of the social scale, its contribution to the nutrition of Ethiopian farm families was negligible. To Ethiopian society and the state it supported, the fate of its livestock was a vital matter. Most important, rinderpest removed this key production factor from Ethiopia's agricultural economy. Its suddenness and virulence traumatized farming communities. Cattle survived in any numbers only in a few places of particularly high altitude and among pastoralists whose transhumance happened to keep them in the right place to avoid the pandemic as it swept through. By the middle of 1889, only one of every ten plow oxen remained to work the fields of Ethiopian farmers, and the loss of this productive force proved devastating.

The literature predominantly projects an image of Ethiopian farmers as committed to an archaic technology, suspicious and fearful of change, and clinging to a past that, with increasing velocity, is ever more remote from modernity. This is a caricature. Ethiopian farmers have good reason to be suspicious of external agents who too often have tried to exploit them or subject them to foreign agendas. Their cultivation practices are far from static and unchanging; have adjusted to transformations in political, economic, and social circumstances; have incorporated new crops; and have shown a market sensitivity often invisible to outsiders. Yet there is perceptible continuity from the late nineteenth into the early twenty-first century. The centrality of the ox-drawn plow to farm life is a case in point.

In the past, as in the present, oxen provide the critically important energy, which launches the agricultural cycle of field preparation, planting, weeding, harvesting, and threshing. To a lesser degree they are also important for planting. Drawing a scratch plow with a single iron tip that undercuts but does not turn over the soil, a team of oxen increases the area a farmer can prepare for planting. One skilled plowman—the role is highly gendered—can manage the team on his own, although on some occasions he might be assisted by other members of his household, who break the larger clods turned up by the plow by hand. Preparation for planting requires multiple passes, ideally after the first rains have fallen and moistened the soil.[89] Although practice varies from crop to crop, planting major cereals is generally carried out by broadcast sowing. Some seed is left as strewn; more of it is covered by another pass with the plow, or by trampling by oxen or people. Oxen contribute little to weeding or harvesting, activities that remain dependent on the availability of human labor unaided by animal or mechanical power. Harvesting is particularly labor intensive and thus, along with weeding, has a limiting effect on how much land it is worthwhile to plow and hence on the value that oxen add to the farm enterprise. Cattle come into play again on the threshing floor, where they tread the harvested grain, which the farmer and neighbors then winnow by hand. Care of the oxen, idle for more than half the year, is managed by younger members of the household and interferes little with other household tasks, although in recent decades it has competed with school attendance.

It is impossible, in the absence of relevant information, to retrospectively calculate the total value the labor of oxen added to Ethiopian households in the course of an agricultural year in the later nineteenth century, and hence to measure the loss that its disappearance entailed. However, some information allows for some grasp of the parameters involved and the complexities of such a calculation. In *Farming Systems in the Tropics* (1976), Hans Ruthenberg compared contemporary tropical agricultural systems and revealed an enormous range in productivity, whether measured by value or volume.[90] In this comparison, fundamental variables were technology—ranging from indigenous hoes to mechanical tractors—and crop. Taking volume of production as the basis for comparison, Ruthenberg showed the variation in selected tropical farming systems, which ranged from the cultivation of pearl millet with indigenous hoe technology in Senegal at 0.00093 tons per human labor hour (thlh) to 0.00348 thlh for upland rice cultivation in the Philippines using machine-drawn tractors. Not surprisingly, the latter were more productive by a factor of over 3.5. Ethiopian plow cultivation of *tef* (at Modjo, 2,200 meters above sea level) and wheat (at Holetta, 2,400 meters above sea level) produced 0.0023 and 0.0021 thlh and came in roughly midway between the Senegalese and Philippines figures, with Modjo farming 2.6 times more productive and Holetta farming 2.3 times more productive according to volume than is the Senegalese example. Conversely, Philippine rice farming was 1.5 times more productive than either Ethiopian case.

A comparison between different farms within Ethiopia, all using the same technology of ox-drawn plows and emphasizing the same crop, gives a further indication of just how variable productivity might be. Ruthenberg provided "farm management data" for three Ethiopian locations—Bako, Holetta, and Modjo, which produced mixed crops, *tef* principal among them. Important factors of variation included differences in altitude, annual rainfall, fallowing practices (and hence soil quality), and proximity to markets. Bako is in the western Shawa highlands. Holetta is nearer Ethiopia's main market, Addis Ababa. Modjo is centrally located within Ethiopia's most developed corridor, between Addis Ababa and Nazreth (now called Adama), and is the most market-oriented of the three. Yield per hectare by weight differs by almost a factor of two, as does yield per hectare by value.

The literature on the history of the plow does not address the extent to which its adoption increased agricultural productivity, other than in general terms.[91] However, there is a literature that takes up theoretical and practical dimensions to the contribution of animal-drawn plows to agricultural productivity, with principal reference to the West African savanna in the later twentieth century. Here government agencies and large companies have promoted the plow, with mixed results.[92] Adoption has been slow and localized, and some farmers abandoned plows after a few years of practice. Where animal traction has been adopted, it seems to allow an increase of 25 percent in the area cultivated per household. Factors affecting adoption are similar to the factors affecting productivity following the adoption and are numerous and complex: length of growing season, soil tilth and fertility, variety of tasks to which the oxen can be turned, maintenance costs for plow animals, and market value of the crops. While the plow makes both land and labor more productive, its adoption brings costs, and the calculation of these costs laid against their putative benefits clearly leads to different decisions. In Ethiopia we are dealing with a situation in which farmers committed themselves to ox-drawn plows more than 2,000 years ago.

The disappearance of most of its draft power disrupted Ethiopian agriculture to a much greater extent than might be indicated by how much it contracted the area under cultivation or decreased the volume of crops a house could produce. It withdrew, overnight in Alaqa Lamma's experience, an essential element to the start of the whole process. Just how short that notice was is seen most clearly in Shawa, where the rinderpest swept through in February, a scant three months before plowing for the major harvest was to take place, and disaster was apparent as early as late July, when Menilek issued his proclamation in response to mounting deaths among his subjects.[93] In some instances, Ethiopia's farmers responded by yoking their horses, mules, and even donkeys to the plow, but these animals were in comparatively short supply—such that in some cases people tried to draw the plows themselves.[94] Moreover, the equines were pack animals, typically lacking the strength of the draft oxen, which in any case have to be trained. Normally, training entailed yoking a young ox with an experienced one, who played an important role in the process. No such trained animals were available any more.

Finally, though not so drastically as the cattle, equines were also affected by rinderpest. Unlike late twentieth-century West African farmers, for whom the technology and techniques remained close to hand, Ethiopian farmers could not so rapidly switch to cultivating with hoes, even were the hoes available. The emperor himself tried to set an example by using a hoe and enjoining his soldiers and then his provincial vassals to follow suit. His chronicler suggests that the response of farmers was passive resignation—they simply became idle.[95] If, indeed, many were resigned, it surely reflected the depth of their trauma, rather than a simple lapse of will.

The role of rinderpest in precipitating generalized famine in Ethiopia in 1889 is clear. Whereas an accumulation of factors, including drought, were at work in the outbreak of famine in Eritrea in 1888, these factors were of less importance elsewhere. Famine followed in the wake of the rinderpest virus as it spread inexorably across the Ethiopian highlands, destroying a vital factor of production.

"Swarms of Locusts and Masses of Caterpillars"

Pest infestations intensified famine conditions. In most cases the pests were locusts, more precisely the desert locust, *Schistocerca gregaria*, which is native to a broad swath of land running from the Atlantic coast of Mauretania through to Pakistan.[96] Twice, in particularly destructive intrusions, the locusts were accompanied by "caterpillars," which quite possibly were army worms.[97] The infestations were limited in place and time. Primarily they affected Eritrea, although they also affected parts of Shawa in September 1891 (specifically Bulga in eastern Shawa) and April 1892.[98] In Eritrea, infestations occurred sometime during late 1888 and early 1889 and recurred in March 1891 and February 1892. These are all likely to have been independent outbreaks. In the accounts of the origins of the Kefu Qan by both Crouzet and Arén, the locusts and caterpillars followed drought and made it more difficult to reestablish normal farming.[99]

Human Disease

Human diseases associated with the great famine include smallpox, dysentery, influenza, typhus and cholera, of which smallpox, influenza, and cholera are most worthy of our attention.[100] None were diagnosed in a clinical, let alone laboratory, setting. Many illnesses were characterized by acute diarrhea, a symptom also produced by the desperate measures to which people resorted in the absence of regular food. The starving cut cattle hides into strips, which they ground into powder and cooked; others scavenged "the dung of mules in the hope of finding grain"; still others ate the meat of diseased cattle and violated food taboos by eating horse and donkey meat; many tried to survive by eating wild grasses and fruits; and some abandoned their homes and took to the road in search of food.[101] Not surprisingly, some of these strategies caused intestinal disorders and weakened people, making them vulnerable to all manner of disturbances and infections.

Several sources recognized the difficulty of distinguishing disorders that were the direct result of famine from those caused by viruses and bacteria. Several observers were trained medical doctors, at least one of whom, Traversi, was called in to treat infected victims; another doctor, Wurtz, carried out an inspection after the fact. Their presence does not seem to have brought precision to disease diagnosis. In some cases, diagnosis was couched with skepticism, in other cases with vagueness, in others with apparent contradiction. In October 1891, writing of an outbreak of cholera in Harar, Stace noted that this might in fact be the "fever of famine"; Wurtz, with somewhat vaguer chronological reference, wrote of "the typhus of famine."[102] Ragazzi, another medical doctor, described rinderpest as a "deadly typhus epidemic," and connected it to a grave form of "dysentery," which was killing people.[103] What all these disorders (with the exception of smallpox) have in common is acute diarrhea. How they could be so readily distinguished, in the absence of firsthand diagnosis or laboratory testing, is unclear.

Even in the absence of famine, correct diagnoses proved elusive. Over twenty years following the end of the Kefu Qan with modern medical institutions in place, in July 1918, the principal Italian representative in Addis Ababa informed the governor of Eritrea that a "fierce variolic epidemic" was raging in the Ethiopian capital. In response to an Ethiopian government request for help, the governor sent 100,000 doses of smallpox vaccination. The epidemic was in fact an episode in the global influenza pandemic. As late as the 1930s, in spite of the attention the Italian administration directed to typhus as a public health issue in Ethiopia, an Amharic–Italian dictionary published in 1940 still confused typhus and influenza.[104] Moreover, the relationship between famine and disease is not easy to sort out. Both smallpox and influenza appear to have been endemic, which means that disease agents already present might flourish in a population whose resistance was lowered by starvation. Ethiopia was probably implicated in two major pandemics: influenza, which ran globally from 1889 to 1892; and cholera in 1892, which killed 10,000 people on the Somali coast.[105] The pandemics may account for influenza in Ethiopia in 1889–90 and cholera in 1892, although there is the possibility that awareness of the pandemics may have led observers to look for diseases that were not actually present.

The first epidemic human disease appeared not on the coast but in Shawa, in July 1889, about five months after the arrival of rinderpest in Shawa. The accounts of the outbreak in 1888–89 in Eritrea of the Great Famine do not mention human disease. That such disease would follow the outbreak of famine is plausible, since it would flourish in the weakened population, but such was not the case in Shawa, where the human epidemic started killing people before the first reports of famine deaths. Emperor Menilek believed that the animal and human epidemics were the same, as he proclaimed in his edict of July 1889: "Now the epidemic has turned to people and has begun to destroy them." This view is also reflected in the account by Alaqa Asma: "The disease which infected the human beings and the animals was the same. It was diarrhoea."[106]

Identifying the human epidemic is challenging, because of the confusion in the earliest accounts, but a close reading suggests that influenza may have been at work. The outbreak of July 1889 was reported by Ragazzi, who identified it as dysentery. This epidemic may have run until October 1890. Its reconstruction is tenuous, complicated by the different terms used to describe it. Two retrospective accounts—by de Lauribar and Wurtz—date a murderous epidemic to the end of 1889 and associate it with an expedition by Menilek to Tigray. It is possible that the army picked up the disease while in Tigray, but there are no independent reports of disease in Tigray at this time, and it is equally possible that the army brought with it microbes, which had first made their presence known in Shawa in July. De Lauribar calls the disease cholera. Wurtz thought it was influenza. Ilg, who was Wurtz's informant, accompanied the emperor, whose expedition foundered on famine and disease—some 20,000 soldiers are said to have been affected.[107] By April 1890, Menilek had withdrawn from Tigray to Dessie in Wallo, where he summoned the Italian doctor, Traversi, to provide medical attention for his army. Traversi reported that the army was "decimated by dysentery" and hunger.[108] What Salimbeni thought was the same disease had by October 1890 "reached virulent proportions" at Addis Ababa.[109] He called it influenza.

What might well have been the same disease was successively called dysentery, cholera, influenza, dysentery again, and influenza again. All these diseases are possibilities. Given the closeness in time of the outbreak of rinderpest and the human epidemic, observers might well have thought that only one disease agent was at work. While they would have been mistaken in thinking so, the error is pardonable. Both rinderpest and influenza were characterized by catarrh, fever, and diarrhea. I suggest that the human epidemic of 1889–90 was influenza, an expression of the global pandemic of these years, and the pandemic, in turn, no doubt much abetted by the new imperialism, which in 1892 was at full flood, forcing new connections across the globe and perhaps bringing new microbes to Ethiopia.

January 1890 saw the first report of smallpox, which, with its characteristic pustules, ought to have been identified with some consistency, although we have seen that this was not always the case. The reports are confined to Eritrea. The outbreak occurred in Massawa. Two reports speak of separate outbreaks in Eritrea in 1892, one in February and one in July, possibly part of the same epidemic.[110] These reports are plausible. Eritrea is a likely place. Massawa was an important port on the Red Sea and nearby Jiddah, through which many thousands of pilgrims passed on the annual Hajj, would have repeatedly functioned as a nodal point in the spread of epidemic disease. The maritime connection between Massawa and Jiddah was active not least because of Ethiopian grain exports to Arabia.

In 1891 we find only two clear references to epidemic disease: both place the incidents in October; one mentions cholera, the other mentions typhus. Stace wrote of cholera, or "starvation fever," being present in Harar, and Traversi wrote of typhus in Shawa.[111] Less clearly, Martini, writing retrospectively of his visit to Eritrea in 1891, spoke of cholera as one of an array of afflictions that had created disaster. Thereafter the sources refer exclusively to cholera.

In 1892 the global cholera pandemic claimed ten thousand deaths on the Somali coast. In April one of two Ethiopian correspondents of Ilg, writing from Deder, west of Harar, said that "the epidemic has not yet abated," and another correspondent, in Addis Ababa, explicitly mentioning *koléra*, advised Ilg not to come via Harar because "they say that the epidemic is raging there."[112] Cholera was reported from the Gondar area in August and from Ankobar in Shawa in September.[113]

By the end of 1892, the famine had passed and with it, the intestinal disorders it had directly caused together with such epidemic diseases for which it had created so fertile a field. Some diseases remained endemic, and Ethiopia remained exposed to pandemics, such as the 1918 influenza, which hit with particular force.[114] Agriculture revived, thanks to adaptation, to the increased availability of plow oxen, and the reduced number of mouths to feed. The state made at best a modest contribution to recovery.

By the end of 1892, the famine had passed and with it, the intestinal disorders it had directly caused together with such epidemic diseases for which it had created so fertile a field. Some diseases remained endemic, and Ethiopia remained exposed to pandemics, such as the 1918 influenza, which hit with particular force.[114] Agriculture revived, thanks to adaptation, to the increased availability of plow oxen, and the reduced number of mouths to feed. The state made at best a modest contribution to recovery.

State, Class, and Famine

State activities increased farmers' vulnerability to famine. This was only modestly offset by relief measures taken by the state, individual nobles, and the charity of churches and monasteries. Warfare was constant, and the plundering by armies stripped farmers of much of the modest surpluses that helped cushion them from environmental stress. Richard Caulk has shown how, there being no salaries, armies lived off the land and preyed on Ethiopian farmers.[115] Reconstruction of the monarchical state, which started with the reign of Tewodros II (1855–68), entailed civil war. Each successive monarch came to power on the strength of his army, and these were not small armies. As indicated, they ranged in number from 10,000 to 100,000, and in some cases entailed numerous camp followers. While Yohannes IV (1871–89) did reestablish royal authority, military conflict continued. Civil war was further complicated by the onset of imperialist encroachments. In 1875 and 1876, Yohannes led successful resistance against Egyptian invasion.[116]

In the short run, victory over the Egyptians gave Yohannes a powerful edge in the struggle to reconstruct the Ethiopian state. In 1878, perhaps at the apogee of his power, he held a religious council at Boru Meda in Wallo at which he imposed a doctrinal unity on the Ethiopian Orthodox Church, which it had lacked for several centuries, and launched a campaign to convert the Muslims of Wallo to Orthodox Christianity.[117] Both unity and conversion entailed violence. At the council, Yohannes also received the submission of Menilek of Shawa and (Ras) Adal of Gojjam. Although these submissions brought some ten years of peace in Yohannes's relations with these two most powerful vassals, it did not end rivalry and conflict between Menilek and Adal, which climaxed at the Battle of Embabo in 1882. Moreover, relative internal peace did not mean external peace. The Egyptian defeats in 1875 and 1876 contributed to the collapse of Egypt's imperial ambitions, but Egypt was replaced by two more vigorous foes—the Italians at Massawa and the Mahdist state in the Sudan.

Following the rise of the Mahdists in the early 1880s, violence flared all along the northern and western borders of Ethiopia. One focus of conflict was Bogos, a territory in Eritrea, which was hotly contested between Yohannes's lieutenant, Ras Alula, and successive authorities at Massawa. Both the Egyptians and Ethiopians plundered and misruled. They also played local rivals against one another, fomenting more armed violence.[118] In 1885, the Italians formally succeeded the Egyptians at Massawa and, like their predecessors, attempted to expand their control into the highlands. Warfare escalated. At the end of January 1887, Alula attacked an Italian advance fort at Saati. Although this attack was repulsed, the next day he successfully destroyed an Italian relief force of some 500 men at Dogali, a victory that doubled the Italian determination to expand. Italian reinforcement of the Saati garrison eventually led the emperor to take matters into his own hands, and in March 1888, Yohannes set siege to the fort.

Meanwhile, the Mahdists, having hastened the withdrawal of Egyptian garrisons on the Ethiopian-Sudanese border, had expanded their control and came into direct conflict with the Ethiopians. This led to battles: at Kufit in the north, in 1885; south of the Blue Nile, in Wallaga, in 1888; and at Matamma, west of Gondar, in 1889. Matamma was the principal site of contact between the Ethiopians and whatever authority held sway in the Sudan. By 1885 the Mahdists had occupied it and used it as a base for raiding. Major battles ensued between the Mahdists and the ruler of Gojjam, now known as Negus Takla Haymanot, in 1887 and 1888; in 1888 the Mahdists sacked Gondar, royal capital of the recent past.[119]

Relations between Yohannes and Menilek had been strained. Initially rivals for the throne, each sought to control Wallo. Tewodros had seen Wallo as a key to reconstructing the kingdom, and he campaigned there with intensity and harshness. The removal of his heavy hand allowed local rivalries to flourish, and Yohannes and Menilek exploited these rivalries to their own ends. Here, as elsewhere in the highlands, armies were the principal instruments of politics. Following Boru Meda there was a degree of peace in the province, Yohannes and Menilek sponsoring the baptism of rival Wallo notables, who became their respective protégés. However, baptism was accompanied in some cases by force, and in 1885 a major rebellion broke out, which was violently suppressed by the armies of Yohannes and Menilek.[120]

As Yohannes's troubles mounted, his principal vassals wavered in their loyalty; in 1888, they formally conspired against him. Yohannes was distracted from his siege of Saati by the Mahdist attack on Gondar, and further distracted from the Mahdists by news of the conspiracy between Takla Haymanot and Menilek. He threatened Shawa and then, starting in September, vented his rage on Takla Haymanot's Gojjam—sacking, pillaging, and burning, ensuring that yet more farmers were brought to misery.[121] Having brutally forced the submission of Takla Haymanot, the emperor turned again to face the Mahdist threat, dying in battle at Matamma on March 9, 1889.

The armies of the 1880s were fighting with advanced contemporary weaponry—Menilek had light artillery—but their systems of supply were minimal. Great hordes, they lived off the land, which meant local farmers and herders were sacked again and again. Menilek repeatedly had to postpone expeditions to Tigray, as the famine-stricken land could not support an army. Moreover, pillage was an instrument of war, as Alula and the Egyptians demonstrated in Bogos, Yohannes and Menilek demonstrated in Wallo, and Yohannes demonstrated in Gojjam.[122] The Kefu Qan was a by-product of the wars of the 1880s. The rinderpest-ridden cattle were imported to feed the Italian garrisons at Massawa and Saati, where they were plundered by Yohannes's besieging army, which then spread the rinderpest in the wake of its march to Gojjam in 1888.

The state was thus deeply complicit in the immiseration of its subjects, but it was also bound, as part of a national moral community, to relieve that misery. And act it did, as did individual noblemen and noblewomen. Churches and monasteries provided such charity as they could. But relief and charity were constrained by the collapse of agricultural production.

Sometime in the first half of 1889, when the scale of the rinderpest epidemic had become clear, Menilek issued an edict telling the people of Shawa what they already knew—that if the cattle disappeared, so would the harvest, without which they would perish. He enjoined his people to turn to their churches and to beg for divine mercy.[123] He also tried to set an example of how to adapt by taking up a hoe and working the land. He distributed hoes to his soldiers, set them to cultivating, and sent hand implements to his erstwhile rival Takla Haymanot and to Ras Mikael, now the most prominent of the converted Wallo nobles, urging them to follow his example.[124] However, by July, as the cattle continued to die and as epidemic disease began to affect the people of Shawa, he issued a second edict, berating his subjects for not having heeded his previous edict, accusing them of not having taken his word seriously and abandoning prayer: "When the animal epidemic was starting, I made a proclamation saying pray to God; but you, oh people, have not taken my words seriously enough and have left off praying. The animals are all dead, and, after you worked the land with a hoe, locusts and worms have devoured your harvest. All this happened because we did not pray enough. Now the epidemic has turned to people and has begun to destroy them." After further injunctions to repentance and prayer, he continued, "I tell you this, since it is fitting that I should be leader of the people whom God has given me, but not because I believe that you will do this. From now on I am no longer responsible for you."[125] In spite of the harsh words, Menilek's chronicler describes him as distressed about holding a coronation celebration in the midst of such misery. But the coronation, which took place on November 3, 1889, was duly celebrated. "Copious food and drinks were distributed . . . for a stretch of eight days." According to Marcus, the principal gesture in recognition of the famine was an amnesty for prisoners.[126]

The starving moved in search of relief—to Addis Ababa, to the houses of notables, to Eritrea and Massawa. In Addis Ababa, "Each day, without interruption," the chronicler says, "the King and Wayzaro Taytu [his wife, the queen] distributed alms to the poor." People arrived from the four corners of the country, as soon as the rivers abated following the end of the rains. Addis Ababa was filled with the starving, large numbers of whom died daily. Traversi credited Menilek and Taytu with doing their best, but resources were too constrained. Public charity was not lacking but was "absolutely inadequate."[127] Wurtz, who shared this view, went on to say that Menilek had "generously opened his granaries" and arranged for caravans to bring rice from the coast.[128] Other nobles provided relief. As to behavior, in addition to setting an example of hand cultivation and doling out relief in person, the emperor is said to have forgone eating beef for three years.[129] There must have been a degree of austerity at the palace, although the chronicler assures us, "while the land was empty, the court, as usual, wanted for nothing."[130] Menilek took actions to prevent hoarding.[131]

These measures were palliatives. Some sources present a reform of 1892 as a structural change intended to relieve the peasants. On October 12, 1892, Menilek introduced a tax, a tithe, which was new to Shawa, although it had been common elsewhere. This was to provide the revenue that would allow him to end the practice of billeting some of his troops on the peasants. These troops had been so billeted for twenty-eight years, and the farmers had already suffered three desperate years carrying them. The measure was introduced several months before it was clear that the famine had passed, so further relief of the peasants may have been one of its objectives.[132] If so, it was very little and very late.

There were no further structural changes in the agricultural lands of the old imperial heartland, although at the local level, with as much as one-third of the population gone, there must have been many adjustments in access to and control of resources. Agrarian Ethiopia emerged from the ordeal with its class structure intact. To be sure, the powerful were not able to wholly escape the famine and diseases, but they surely fared better than did farming people. Their surpluses were larger, and they controlled the institutions of violence, which afforded them a disproportionate share of what production there was. Nevertheless, they also suffered. We have seen how at the end of 1891 the women and children accompanying the "Chiefs" of Tigray were "haggard, emaciated and in a wretched condition."[133] Further evidence indicates that the lack of food forced some Shawan military leaders to break up their units, sending the soldiers in search of food and others to lead their troops in raiding those pockets of relative well-being, which had survived the early onslaught.[134]

In the larger picture, the Kefu Qan may have affected the ongoing expansion of the empire and Ethiopia's relations with imperialist Europe. Territorial expansion was a process dating back to the first half of the nineteenth century. It accelerated during the era of imperialism, although not at a uniform pace, the Kefu Qan perhaps acting as a temporary brake. The Ethiopian ruling classes still commanded the most concentrated military resources in their immediate region, which would further have increased their ability to dominate and incorporate neighboring peoples, who had also suffered from famine and disease. The Kefu Qan clearly delayed Ethiopia's confrontation with Italian imperialism, until the Battle of Adwa in March 1896, at which the two armies, equally short of supplies, finally came to grips.[135] The delay may have affected the Ethiopian ability to capitalize on their battlefield victory and equally have allowed the Italians to consolidate their position on the Eritrean plateau. The upshot was that the victorious Ethiopians were unable totally to expel the Italians, whose colony of Eritrea became, through the next century, a significant participant in regional politics.

Reflections on the Kefu Qan

In a number of ways the great famine of 1888–92 foreshadowed the famines of the 1970s and 1980s. In other ways it was unique, the product of a moment, of an unrepeatable chain of events. Oddly, one point of similarity is the uncertainty of the scale of the tragedy. The Kefu Qan was said to have claimed something like one-third of Ethiopia's population. The famine of 1973 is said to have claimed 100,000 lives, while in 1984–85, it is said that 500,000 died. Both are estimates, which perhaps tell as much about the intensity of the impact of events on observers as they do about the scale of those events. This account of the famine has pushed the available evidence to its limits. In so doing, it may have been misleading. Obviously Ethiopia survived the famine and its agriculture revived, as did human population, which boomed in the later half of the twentieth century. Throughout the years from 1888 to 1892, there were indications of limits to the scope of the famine.

Food continued to be produced. Our best evidence comes from Eritrea and Shawa and, in spite of the horror of the accounts of the height of the famine, many survived. In 1891 Martini believed that the starving hordes on the notorious plane of Otumlo, outside Massawa, had come from Tigray and that economic stimulus provided by the establishment of the colony of Eritrea, formally proclaimed in 1890, had created employment. In Hamasen, the highland district of Eritrea, closest to Massawa, he claimed, with imperial self-satisfaction, "there was bread for everyone."[136] At the entrepôt centers of Massawa and Matamma, food was to be had, for a price. The Italian Geographical Society's station at Lat Marafeya, thanks presumably to its access to external resources, maintained more or less regular agricultural production.[137] Although the areas immediately around Harar, populated by the Ittu Oromo, were suffering grievous famine, the Ethiopians (who had only recently taken over the area from the Egyptians) suffered less if at all, since according to Mashcov, they had also occupied the Ogaden region, to the east of Harar, "rich in grain."[138] In August 1889, Rimbaud wrote of the harvest that coffee had failed but sorghum was "mediocre." In 1891 Crouzet described an Easter feast in Bogos, where goats, sorghum *enjera*, and mead were consumed in plenty. As late as June 1892, Picard, writing from Eritrea, could observe that the misery continued, the poor were forced into exile, and even the rich had become too poor to buy enough food for the year, but nevertheless, people still had cows to sell, along with goats and mules.[139]

In the nineteenth and the later twentieth century, internal and external forces combined with deadly effect. But the combination was distinct to the moment. The Kefu Qan was the product of the high imperial era of the end of the nineteenth century. Without rapidly developing global networks, the advent of rinderpest would have been postponed indefinitely, if, it would strike at all. Influenza and cholera pandemics would have unfolded very differently and at different times. Although European imperialism is an inescapable dimension of the great famine, the famine was not the product of a capitalist transformation in the forces or relations of production or of the political economy of a state modernizing by internalizing capitalism. Menilek's state was thoroughly precapitalist in its internal economy and its external commercial relations, which increased steadily throughout the nineteenth century, continued patterns of supply reaching well into the past. The principal exports—ivory, civet, gold, hides, and coffee—were obtained within the framework of long-established social relations and practices.[140] The political economy of the famine was not capitalist.

The decisive environmental factors at work were microbes, and, more modestly, climate. Some of the microbes were long established, others were new. The rinderpest virus brought about a collapse in agricultural production because of the centrality of cattle to cultivation. Other microbes flourished in the human populations weakened by the lack of nourishment. Microbes played at best a modest role in the great famines of the later twentieth century, although to be sure, they will always take advantage of the reduced resistance brought on by starvation, while climate came increasingly to the fore. Politics, the economy, and Ethiopia's global setting were transformed. The social relations and forces of production for the great majority of Ethiopia's highland farmers changed surprisingly little.

2

Famine in Ethiopia, 1973–1974 and 1984–1985

FAMINE WAS NOT MUCH ON THE MINDS of the political and intellectual elites in Ethiopia in the early 1970s. A colleague in the History Department at Haile Selassie I University once remarked that famine was not a feature of twentieth-century Ethiopian history the way that it was of India and China.[1] To be sure, we knew of chronic food shortage and emigration from one small part of northern Ethiopia, the region of Wag and Lasta, but this was geographically limited, and there had been the Kefu Qan, but they were the product of an unfortunate historical conjunction—the introduction of rinderpest to northeast Africa. In the late 1960s and early 1970s, I traveled a good deal around northern Ethiopia and was not confronted by famine and food deprivation—until May and June 1973, when I journeyed with a few friends north from Addis Ababa to visit the monastery of Dabra Damo in Tigray. We found the road through northern Shawa and southern Wallo to be thronged with people, desperately hungry, begging for food from passersby. They were famine victims and harbingers of misfortunes to come.[2]

We had been alerted to the possibility of famine by some of our Ethiopian colleagues at the university, two of whom—Mesfin Wolde Mariam and Abraham Demoz—heard rumors that large numbers of people from Wallo, driven by hunger, had come to Addis Ababa in February 1973, on their way to seek work in the coffee-growing areas of southern Ethiopia; they visited them and learned of wide-scale food shortages. Rebuffed in their attempts to get the government's attention, the two scholars determined to learn more of the situation. In the middle of April, joined by

Getatchew Haile and Alula Abate, they flew to Wallo. In the course of a day's local travel, they gathered photographs and recorded searing evidence of starvation, particularly among women and children. They brought this evidence back to Addis Ababa, where they arranged an exhibition at the university.[3] I do not recall visiting this exhibition, but I remember the relief campaign my colleagues mounted within the university. It took a direct encounter with the starving to make me fully aware of their plight, and even that was not enough for at least one of my traveling companions, who complained of the nuisance all these people on the road created.

Famine, 1973–1974

The exposé by Ethiopian faculty members proved to be the start of a growing awareness in the nation and internationally of what became commonly known as the Wallo famine of 1973–74 in which perhaps some 100,000 people died.[4] Estimates of mortality from famine range from 50,000 to over 400,000, and the area to which they refer ranges from Wallo to the entire country. In his sober account of this famine, Amartya Sen settled on the number of 100,000, which may be the best we can do.[5] This awareness swelled and became global in 1984, when famine recurred with much greater intensity, with Wallo again at its epicenter. But 1973 was not Ethiopia's first year of famine, not even in the latter half of the twentieth century. Mesfin established that this story began in 1958 with famine in Tigray province immediately to the north of Wallo.[6] Nor did the famine we became aware of in 1973 start in that year. It began earlier with irregular rainfall in 1971 and 1972. The *keremt* rains, expected in June, July, and August, failed, bringing failure of the principal harvest and famine, which was then compounded by the failure of the *balg* rains of February and March 1973 and the lesser harvest that should have followed. Nor in 1973 was famine confined to Wallo. From the start it affected Tigray and northern Shawa; by 1974, it had spread to the highlands of Harar and neighboring lowlands and to Gamu Gofa in the southern Rift Valley. It was not confined to Ethiopia; rather it affected the entire Sahelian region, a band running east–west across Africa, ecologically transitional between the Sahara and the savanna lands to its south.

Awareness within Ethiopia grew slowly. The government controlled the media, which were confined to Addis Ababa—there were no independent media outlets in Wallo, Tigray, or northern Shawa. It was determined to suppress all information about drought and famine. No one within the government wanted to be the bearer of bad news. As Mesfin noted, "The whole bureaucracy had developed the art of sifting information in order to make it acceptable to higher authorities. The higher the information moved, the more it lost its original, crude and unpleasant conformity with the reality."[7] Communication between local officials in Wallo and the authorities in Dessie was erratic, as was communication between Dessie and Addis Ababa. Nevertheless, by early 1973 the national Ministries of the Interior and of Agriculture and the elected parliamentary representatives from Wallo had plenty of detailed information about drought and famine.

In November 1972, a production survey by the Ministry of Agriculture indicated that as many as 250,000 people in Wallo faced major food shortages and that contiguous areas to the north and south along the edge of the Rift Valley—Shawa, Tigray, and Eritrea further north—also faced serious shortages. The report focused on aggregate production in the districts making up Ethiopia's various provinces. Wallo reported that 90 percent of its districts were "below" or "substantially below" normal yields. Hararghe was the next most challenged province, 39 percent of its districts reporting below or substantially below normal production. The report noted, "The most serious problem is centred in the lowland areas east of the escarpment road in Tigrai and Wello and the lowlands of northern Shewa."[8] This report was not published until March 1973.[9] At the end of that month, the governor of Wallo had submitted a detailed report to the Ministry of the Interior, as had a government fact-finding team. The Wallo parliamentarians attempted to raise the issue. Following their trip to Wallo, the faculty members from the university tried to spread information and pressed for government action.

What was obvious to every traveler on the main highway from Addis Ababa to the north was dismissed by the authorities as a "fabrication" and "lies," although they did take begrudging steps to address the famine.[10] In April, the prime minister reactivated a National Emergency Relief Committee (NERC), which had been founded in 1971, and charged it with mobilizing funds and gathering supplies. It was composed of cabinet ministers and other government officials, a group complicit in blocking dissemination of information and effective action. The NERC contacted international agencies and embassies in Addis Ababa, but the government still tried to minimize the scale of the looming disaster. The NERC did "not see the need" for the large-scale, public meeting that donors called for to better coordinate relief efforts. Government reluctance to admit the existence of famine was reinforced by the impending celebrations in Addis Ababa to celebrate the tenth anniversary of the founding of the Organization of African Unity.[11] The first public official hint that something was awry came at the end of April, when the government newspaper, *Addis Zaman*, reported that relief had been sent to Wallo and northern Shawa and that the government had requested international relief assistance three days later. Still, neither government nor the international community had fully grasped the dimensions of the problem.[12]

Following the failure of the spring rains in 1973, farmers and herders on the lower levels of the western escarpment of the Rift Valley flocked to towns along the main north–south highway. By June 1973 relief camps had been established at strategic points along the road from Tigray in the north through the main towns in Wallo to Shawa in the south. Concentrating destitute people in this way allowed epidemic diseases to flourish. By August, more than 60,000 people had crowded into the camps. Another 200,000 people were looking for relief. August marked the high water mark of distress. The situation eased when good rains in Wallo in *keremt* 1973 raised the prospect of a good harvest, which encouraged many to return to their farms and induced the holders of surplus grain to release it to the market.[13]

By the end of September, in Tigray and Wallo "the starvation crisis was over," and by October the population of the camps had dropped to 15,000.[14] Ironically, relief then flooded in, boosted by *The Unknown Famine*, a televised exposé of the famine in Wallo by Jonathan Dimbleby of Thames Television in the United Kingdom. First broadcast on October 18, 1973, it was widely disseminated and proved to be the first of what a decade later was a succession of horrifying images of starvation in Ethiopia.[15] Up until November 1973, the government had distributed a meager 12,000 metric tons of grain as relief food, of which 6,500 metric tons went to Wallo. In that same month, the government issued a statement blaming the delay in responding to the famine to a "cover-up," which it blamed primarily on Solomon Abraham, acting governor of Wallo.[16] The government also reported that after "carefully taking statistics from registrations at relief camps and interviewing awraja governors, estimated that in five provinces—Tigre, Wollo, Shoa, Hararghe, Begemdir, and Gemu Gofa—some 1,098,300 adults and 470,700 children were starving."[17] Over the following year to December 1974, Ethiopia received over 126,000 metric tons of foreign relief grain. This time 70 percent went to Wallo and Tigray, while Harar, to which the famine had already moved, got only 8 percent.

The events of 1973 cast a cruel light on government in Ethiopia.[18] The modern Ethiopian state was an autocracy whose recent origins were in the second half of the nineteenth century. Modernization came slowly, sometimes begrudgingly, but by 1970 Ethiopia had a skeletal network of all-weather roads, a telecommunications system, a university, and a respected airline. Government revenue came primarily from customs and exports. Following the end of the Italian occupation of 1936–41 and Haile Selassie's return to power, a series of proclamations in 1942, 1943, and 1944 laid new fiscal foundations for the state and undermined the political role of the landed nobility, who had been the principal intermediaries between the Addis Ababa government and rural producers. The Ethiopian government was disengaging from Ethiopian society. Increasingly, it relied on indirect taxes for revenue—customs and export levies. As late as 1972, direct taxes provided only 27 percent of total government revenue.[19] Despite its introduction of income, health, and education taxes, it did not provide many services to its subjects. Ethiopian health and education benefits compared unfavorably with those in many other African countries. In 1968–69 Ethiopian secondary schools enrolled only 1.9 percent of the relevant age cohort.[20]

Emperor Menilek had introduced a ministerial form of government in 1907, and Haile Selassie further developed it.[21] Ministries eventually covered agriculture, commerce, defense, education, finance, foreign affairs, interior, and justice. Particularly Ethiopian was the Ministry of the Pen, a formalization of the centuries-old office of *sahafi ta'ezaz*, the "writer of the orders," or recorder of the imperial edicts. Ethiopia was still an autocracy, ministers were dependent on and reported directly to the emperor, in spite of the new constitution of 1955 introducing a parliament of popularly elected representatives. Government still revolved around Haile Selassie's court, where all real conflicts were eventually resolved, if at all. The emperor still controlled the government with instincts honed in the courts of the late-nineteenth-century nobility—nurturing rivalry and intrigue among subordinates and carefully constructing offsetting centers of power and influence. Characteristic here was the 1959 creation of a short-lived Private Cabinet parallel to the publicly recognized Council of Ministers. Haile Selassie was the center of power, but he was aging—he celebrated his eightieth birthday in 1972—and running out of energy. He directed his attention increasingly to foreign (rather than domestic) affairs.

His handling of the famine indicated either callousness or a remarkable degree of insulation, high enough to survive visits to Wallo in November

1972 and November 1973. On the latter visit he issued an edict that land, which desperate peasants had sold to buy food, should be sold back to them at the original price plus interest, which, as Shepherd notes, was never implemented. In an interview with the Italian journalist Ariana Fallaci, which was published in the *Chicago Tribune* on June 24, 1973, the emperor is reported as saying:

> Rich and poor have always existed and always will. Why? Because there are those that work and those that don't, those that wish to earn their living and those that prefer to do nothing. Those that work, that want to work, are not poor. . . .
>
> We have said that whoever doesn't work because he doesn't want to is poor. We have said wealth has to be gained thru [*sic*] hard work. We have said those who don't work starve.[22]

The famine crisis of 1972–74 revealed the inability of his government to act and played a role in the overthrow of the imperial state in September 1974.

Revolution, 1974–1991

The famine of 1973 was followed by the outbreak of revolution in Ethiopia in 1974. The connection between the two events was more tenuous than sometimes believed.[23] The famine had given warning that the Ethiopian government was dysfunctional and sclerotic. So it should have been no surprise when it proved unable to contain unrest, the first indication of which was a military rebellion in January in Negele in southern Ethiopia sparked by miserable conditions of service. Rebellion spread through the armed forces, first to the Dabra Zayt Air Force Base outside Addis Ababa and then to the army garrison in Asmara in Eritrea, where once again conditions of service were harsh. At this stage, developments in the military took place out of the public eye, which was not the case with a national teachers' strike in mid-February, directed at resolving multiyear negotiations on pay and conditions and against a proposed curriculum overhaul that would stress vocational rather than academic skills. This was followed closely by a strike by Addis Ababa taxi drivers against a 50 percent rise in the cost of fuel without any compensatory rise in the fares they were allowed to charge. This increase was set by the Ethiopian government in response to the Arab oil embargo of September 1973, which followed the 1973 Arab–Israeli War. These strikes sparked a wave of protests and demonstrations, which by the end of February led to the dismissal and replacement of the prime minister and his cabinet.

From then on, discontent swelled in Ethiopia's cities and towns as civil servants, industrial workers, the army, and students protested. The Ottaways, who witnessed these events firsthand, describe the period from March to June 1974 as one of "indescribable chaos."[24] Military protest became more public. Air force helicopters from Dabra Zayt dropped leaflets over Addis Ababa. The principal demands of the demonstrations were the removal of corrupt leaders or government officials and the right to organize. The countryside was not involved in these events at this time, and for months to come; the activity was confined to Ethiopia's principal towns.[25] The chaos was slowly overcome following the formation at the end of June of a committee to coordinate political action within the armed forces, outside the regular structures of command. This became known as the Darg, which deposed Emperor Haile Selassie at the beginning of the Ethiopian New Year in September.

Inflation is cited by some authorities as contributing to urban unrest.[26] The steep rise in the cost of fuel generated inflationary pressure as it worked its way through the rest of the economy. Lefort's figures indicate that through 1973 the general retail food price index rose by 24.4 percent and the cereal price index rose 43.9 percent. According to Gilkes, for the first three months of 1974 the rate of inflation in retail prices rose 80 percent per quarter.[27] To what extent might drought-induced food shortages have contributed to national inflation?

Despite the lack of directly comparable data from one year to the next, the inflationary possibilities of crop failure have received close attention. Sen reviewed the evidence for a shortfall in the availability of food grains in 1973, the year when, following the poor harvests of late 1972 and the first quarter of 1973, it would be most acute. Using figures from the 1972–73 survey of grain production by the Ethiopian Ministry of Agriculture, he argued that the fall-off in cereal production in 1973 was less than 6 percent measured against 1971–72 figures, and there was "very little evidence of a dramatic decline in food availability in Ethiopia coinciding with the famine." He noted that the national bank calculated that overall there had been a "modest increase in agricultural output," when the famine year was compared with preceding years. Sen cited the UN Food and Agricultural Organization as reporting that in 1973 Ethiopian national calorie consumption showed "no significant diminution and possibly a sizeable rise over the

TABLE 2.1. Ethiopian food grain production (in metric tons)

Year	*Barley*	*Tef*	*Wheat*
1972	1,564,000	1,380,000	922,500
1974–95	892,000	804,400	484,000
% change	-43.0	-41.7	-47.5

Source: Mehari and Vahlquist, "Famine in Ethiopia," 198; for 1972, FAO Country Report; for 1974–75, World Bank Report.

preceding years."[28] By contrast, Mehari and Dahlquist, who draw figures from a slightly different period—1972 versus 1974–75—and possibly from different data sources (an FAO Ethiopian Country Report for 1972 and a World Bank Report for 1974–75) do suggest a much larger production fall-off. They cite national grain production in 1972 to have been 6,208,900 metric tons, whereas production for 1974–75 had declined by almost 27 percent to 4,544,000 metric tons. Table 2.1 presents the fall-off in production of three important cereal crops.

There were differences in the prices and availability of grain between the national scene and the regional one in the immediately affected areas. Holt and Seaman appear to be a leading source here. They gathered food prices in markets in northern Wallo and the regional capital, Dessie. They found that prices in the markets of north Wallo were about five times the normal level for that time of the agricultural year, whereas in Dessie they were about half that, "but still much higher than the seasonal norm." They also heard "little clear evidence" of food shortages in the Addis Ababa markets. By the end of 1973, they found that prices in Dessie and in the north Wallo markets had returned to within 15 percent of "the recent pre-drought level." Crop failure did indeed drive food prices up, but only within the limited confines of Wallo, and by implication Tigray, and for a limited time. As far as availability of grain was concerned, Holt and Seaman agree with Sen that grain production in Wallo had probably been sufficient to prevent large-scale starvation.[29]

The Darg used images of royal and aristocratic indifference to popular suffering in the propaganda, which immediately preceded the deposition of the emperor. It broadcast a TV program that interspersed images of starving country folk with images of the emperor feeding his dogs off a silver tray.[30] Famine and the publicity surrounding it helped strip the emperor of the mystique that surrounded monarchy in Ethiopia (at least in the eyes of those with access to television). Meanwhile, the transitional government of Prime Minister Endalkachew Makonnen created the Relief and Rehabilitation Commission (RRC), under the direction of Shimelis Adugna, a man of energy and integrity. By June the commission had a staff of over thirty. Even though the crisis of famine had passed, by April 1974 forty different relief organizations were at work in Wallo and Tigray, although by now the famine had moved on to southern Ethiopia, where many thousands of people especially in Hararghe and neighboring regions faced famine.[31]

On September 12, 1974, Haile Selassie was deposed by the Darg. Originally patriotic and reformist—its first slogan, put forward in July 1974, was *Ityopya teqdem*, "Let Ethiopia be first"—the Darg was steadily pushed to the left by the student movement, which had arisen in the late 1960s on the campus of what was then Haile Selassie I University. The students were dismayed by Ethiopia's poverty and backwardness, not least in the context of Africa, and were inspired by the international revolutionary movements of the 1960s. Their interest in Marxist-Leninist ideology was fed by Ethiopian student diasporas in Europe and the United States. Already by 1965, the students had protested under the slogan "Land to the Tiller." From 1970 onward they were more radicalized, Marxist-Leninist ideology became hegemonic in the movement, and as the government increasingly tried to contain the university students, radical activism spread rapidly through the secondary schools in Addis Ababa and the main provincial towns.[32] Almost from the beginning, the student movement was at odds with the Darg and insisted on a transition to a civilian government. To rid itself of this thorn in its flesh, the Darg organized a Development through Cooperation Campaign and, at the end of December, sent the students to the countryside to promote the revolution, encourage literacy, and explain the forthcoming land reform program to the peasants.[33] The campaign became commonly known by its Amharic label, *zamacha*, the same word for a military exercise. The arrival of the campaigners brought the revolution to the countryside and precipitated struggles over land in many parts of the country, but not in famine-prone northern Shawa, Wallo, and Tigray.

As events unfolded from February 1974 on, the land question came to the fore in Addis Ababa. The student movement had kept the issue alive under the emperor, and it soon became the subject of heated debates in the parliament. There was no popular pressure from the countryside—no uprisings, no *jacqueries*. Land reform had officially been on the national agenda since 1961, when, under pressure from international donor agencies, the government created what by 1966 had become the Ministry of Land Reform and Administration.[34] One goal was ostensibly the improvement of peasant agriculture and through it of national agricultural productivity, another was greater equity in rural society, and yet another, perhaps held by a minority, was the creation of a land tenure system more congenial to capitalist investment. The ministry, for all of the good intentions of many of its employees, was window dressing on the government's part. The famine changed things, and land reform became the rallying cry of those who sought to remedy what they saw as the failures of peasant agriculture, accelerate agricultural development, and overcome famine vulnerability. The reformist administration of Endalkachew Makonnen commissioned the Ministry of Land Reform to draft a proposal for tenure reform. The ministry ended up making its recommendations to the Darg. Concern with the land and access to shortly became infused with environmental rhetoric.

Out of this came, on March 4, 1975, a radical and sweeping measure, Proclamation 31, "A Proclamation to Provide for the Public Ownership of Rural Lands," which nationalized all agricultural land in the country and provided for its administration through peasant associations.[35] Farmers no longer owned their land, although the legislation recognized their rights to have access to it and benefit from its usufruct. They were no longer allowed to rent land or hire agricultural labor. The legislation mandated that the upper limit of holdings be ten hectares, but no holding ever approached this. A further proclamation in December 1975, "A Proclamation to Provide for the Organization and Consolidation of Peasant Associations," clarified the functioning of the associations and gave them the more explicit charge of fostering "socialist" agriculture.[36]

The arrival of the student campaigners precipitated conflicts over land, leading to the overthrow of landlords and the transformation of local administrations in much of the country, especially in the south. Although this did not happen in Wallo or Tigray, the land reform of 1975 did have radical effects and established the conditions under which most farmers still hold land in the region. It did this by a thorough redistribution of existing holdings, a provision for future redistributions, and setting parameters that strictly limited landholdings. The land reform resulted in a compression of land holdings within the peasant associations; the difference between the largest and smallest holdings was now much narrower. The reform did not directly affect the production process, which was still controlled by households working individual holdings, nor did it affect the distribution of other assets, like implements, oxen, or other livestock. It made movement between associations very difficult, although this may have affected men more than it did women, despite the recognition of the rights of women to hold land. In practice men were recognized as heads of households and women, more often than men, crossed association boundaries to marry.

The peasant associations, which were to be no larger than 800 hectares in size, became the primary administrative unit in the countryside, the government's main tool of rural control, and the site of local political conflict.[37] Although the adjustment of details was ongoing, the steady population growth meant an equally steady decline in the size of holdings. The fall of the Darg in 1991 had little effect on this process, at least in Wallo, since the succeeding government, led by the Ethiopian People's Revolutionary Democratic Front, kept the same legal framework, within which they organized a second major land redistribution in 1997.

The Darg adopted a Russian model of agricultural production through peasant cooperatives and state farms. The state farms were nationalized large-scale farms and plantations of the imperial era. In 1997 I interviewed 120 farmers from thirteen different peasant associations in three different *warada*—one in Wallo and two in Gondar. None had state farms in them. The Peasant Association Consolidation legislation anticipated the creation of peasant cooperatives. Although these were influential—they were sited on the most favorable agricultural lands and received preferential investment—they affected only a few farmers. Only one of the thirteen peasant associations in which my informants had lived had been organized into a cooperative. Between them, the state farms and farmers' cooperatives never produced more than 10 percent of Ethiopia's annual crop and they did this with less efficiency than the individual households organized through the peasant associations. The land reform was resoundingly successful in achieving its goal of greater equity in the Ethiopian countryside. But it failed to reach the goals of increasing productivity of peasant agriculture, paving the way for capital investment, and reducing famine vulnerability.

In addition to the perhaps unintended way the land reform and peasant associations restricted farmers' options, many more restrictions were on their way. The Darg levied land-use and agricultural income taxes, and local officials regularly demanded "contributions" to numerous special campaigns and in support of the Revolutionary Ethiopian Youth Association, the Revolutionary Ethiopian Women's Association, and of a host of local projects. The Darg also levied compulsory unpaid labor. It created the

Agricultural Marketing Corporation, which had the authority to set prices paid to farmers and determine the amount of produce farmers were to sell. On the grounds that "frequent" attendance by farmers at markets was adversely affecting their productivity, the government determined that all markets were to be held just one day a week, Saturday. Historically, farmers had availed themselves of the opportunities offered by a network of markets, which rotated through the days of the week. This allowed most farmers access to several different markets, which in turn provided several sources of information and gossip and a chance to compare prices. Still more onerous and resented was the military draft, announced by Mengistu in his May Day speech in 1981 and instituted by the National Military Service Proclamation issued in May 1983.[38] Farmers described the draft as the "tax on children."

Complicating issues of economic development and later famine was civil war in Tigray and Eritrea. The Darg was never able to capture all the dissident energy unleashed in Ethiopia in 1974. The student movement of the Haile Selassie years proved to have been a fertile incubator of would-be revolutionaries. Two parties led the civilian opposition to military rule: the Ethiopian People's Revolutionary Party (EPRP), made up largely of students from what was now Addis Ababa University and Ethiopian high schools; and the All-Ethiopia Socialist Movement, composed largely of students educated in Europe and North America. The latter drifted ever closer to the Darg. Both were crushed during the Red Terror of 1976–77. The EPRP regrouped as a guerrilla movement in Tigray and Gondar but was again crushed by the insurgent Tigray People's Liberation Front (TPLF). The two northern provinces of Tigray and Eritrea supported armed groups bitterly opposed to the Darg. The Eritrean People's Liberation Front (EPLF) traced its origins to the late 1950s. It had burgeoned in the 1960s and, with the outbreak of revolution in Addis Ababa, had seized most of Eritrea by 1975. These gains were reversed in the Darg's Red Star campaign, which in 1982 succeeded in recovering control of much of Eritrea but failed to destroy the EPLF. Under the shadow of the EPLF, Tigrayans steadily deepened their roots in the countryside and, from the early 1980s onward, steadily increased its presence in Tigray.[39] For the peoples of Tigray and Eritrea, the decade once again added famine to the affliction of war.

Famine, Environment, and Development

The famine of 1973–74 established a link between famine and environmental degradation in the minds of the public, politicians, and relief and development agencies. In 1972 the United Nations sponsored the Stockholm Conference on the Human Environment, the first step in creating a web of international environmental institutions. The conference gave birth to the United Nations Environment Programme, headquartered in Nairobi, the first major UN agency to be based in Africa. Meanwhile famine broke out in Africa, affecting the entire Sahelian belt. International attention was first directed to the West African Sahel, giving rise in 1973 to the United Nations Sudano-Sahelian Office (UNSO), the Office to Combat Desertification and Drought of the United Nations Development Programme. These institutions formed the nexus within which global thinking about the environment, economic development, and vulnerability to famine developed. Development agencies such as the World Bank, the US Agency for International Development, and the Swedish International Development Cooperation Agency directed their aid to projects informed by this thinking, as did such nongovernmental agencies as the International Red Cross and the World Wildlife Fund.[40]

Alemneh Dejene identifies 1978 as a key time when the circulation in Addis Ababa of "Reversing Africa's Decline," a paper by Lester Brown and Edward Wolfe, crystallized thinking about the environment and development in Ethiopia. Brown and Wolfe perceived processes of "deterioration" and "degradation" to be at work in the Sahel famine, manifestations of the breakdown "in the relationship between people and their natural support systems." Their paper called for "a long-term strategy of environmental restoration," which would include "slowing population growth, conserving soils, restoring forests and woodlands."[41] It spoke to a void in the thinking of Ethiopian officials and intellectuals, who before 1974 had lacked a coherent approach to "soil conservation or natural resources management," and provided a script for action. This dovetailed with the actions of those aid agencies that had redirected their efforts and relief grain to environmental rehabilitation in the aftermath of the 1972–73 crisis.[42] It focused attention on soil erosion as central to the discussion about farming and the environment in Ethiopia.[43]

This moment represented the confluence of global institutions with their embedded script and local academic and political elites and has continued to shape thinking about the environment in Ethiopia to the present. The paper by Brown and Wolfe had been published by the Washington-based Worldwatch Institute, which Brown had founded in 1974. In 1981 the Ethiopian Ministry of Agriculture and the University of Bern jointly sponsored the creation of a soil conservation research project led by Swiss soil scientist Hans Hurni.[44] This was followed in 1983 by the founding of the Ethiopian Highland Reclamation Study (EHRS), jointly sponsored by the Ethiopian government and the UN Food and Agricultural Organization, supported by funding from the World Bank. The EHRS was a large-scale survey shaped by the concerns articulated by Brown and Wolf and institutionalized in the United Nations Environment Programme. When famine broke out again in 1984, there was a formidable arsenal of institutions and explanations ready to hand.

Famine, 1984–1985

In 1984 Wallo was again at the famine epicenter.[45] This famine affected a wider area and many more people than its predecessor, and it followed a pattern laid out in 1973 and 1974. Prolonged drought appears as a primary cause. It affected many of the same geographical areas. Once again relief was distributed through large camps on the main roads. At a critical moment relief was delayed by the Ethiopian government directing its attention elsewhere. The famine unfolded slowly, painfully, and clearly in the sight of the international community.

In March 1981 the *New York Times* reported that famine loomed in Africa, threatening 150 million people. The story covered a wide swath of the continent running across the Sahel and southward from the Horn through East and Central Africa. Drought was a factor.[46] In 1982 the story picked up again—a recitation of intensifying drought and drought-related famine and steadily growing numbers of threatened and affected victims. As the famine was barely contained chaos, there was uncertainty about the numbers.

In October 1982 the Ethiopian government, anticipating a shortfall in grain production of 250,000 tons, circulated a request to the foreign embassies in Addis Ababa for food assistance. That same month, the Save the Children Fund opened a feeding center for malnourished children in Koram, northern Wallo. Some 300 to 400 children were served.[47] In March 1983 the *Christian Science Monitor* reported that drought had brought crop failure and the prospect of starvation to up to 450,000 people in Eritrea, Tigray, Wallo, and Gondar. That same month Ethiopia declared a state of emergency. The World Food Program and Western countries responded with promises of some 100,000 tons, of which little was delivered.[48] Food aid became a political football. The United States refused relief on the grounds that Ethiopia was a communist country. The Europeans were convinced that the RRC was inflating the numbers. Addis Ababa and the Tigray insurgents accused each other of manipulating food aid for political advantage.[49]

The situation worsened. April 1983 brought news of a surge of Ethiopian refugees into neighboring Sudan, many of them fleeing famine and war, and of a need for 300,000 tons of relief food.[50] The number of Ethiopians threatened by drought was now estimated to range from 2 to 5 million. By July there was an increase in international food aid; despite logistical problems, the food aid was moving into the affected areas. The US government seemed more disposed to provide assistance. In Tigray, however, the provision of relief was fraught by continuing fighting between the Ethiopian army and the Tigray Peoples' Liberation Front, which now controlled much of the province.[51] By October civil war in Eritrea joined the ongoing conflict in Tigray to stymy relief efforts in northern Ethiopia.[52] Political charges and countercharges between the government and its insurrectionary opponents were not the only source of controversy surrounding famine relief. In December 1983 the *Sunday Times* of London alleged "diversion and cover-up" by the Ethiopian government in a story headlined "How 15,000 tons of famine aid went 'missing.'" Charges of misdirection of food aid continued to be made.[53]

Famine continued slowly to build into 1984. Military conflict persisted at a high level in the north, and by February drought hit a new area, Sidamo in the south. The United States was providing modest "humanitarian" assistance, which it channeled through NGOs. Ironically in light of subsequent developments, the Associated Press reported, "By most accounts, Ethiopia was far better prepared for the current famine than the one in the early 1970s and has managed to forestall mass starvation."[54] In March Dawit Wolde Giorgis, the new director of the RRC, unsuccessfully circulated an appeal to key sectors of the Ethiopian government asking for urgent action on famine relief. In May, he traveled to New York and appealed to the UN General Assembly for "a generous and concerted action by the international community" to forestall imminent famine, which "a drought of unprecedented severity" threatened.[55]

The press reports reflect a constantly shifting and worsening situation. In July *The Guardian* cited the charity Oxfam to the effect that the Ethiopian government had now requested 900,000 tons of food assistance, which potential donors had "pared down to 125,000 tons." The village of Korem, which in normal times had a population of a few thousand, was now said to be an enormous relief settlement for 365,000 famine victims. Four months later, the *Baltimore Sun*, reporting from Alamata, a few miles from Korem, put the Ethiopian government request at 600,000 tons and the Western donors pledge at 160,000 tons "so far." It claimed that there were as many as 100,000 famine refugees at Korem with "at least 200,000 more famine victims within a 60-mile radius of the camp."[56]

In June, Mengistu mentioned in a speech that "Ethiopia is currently suffering from severe drought," and went on to assure his listeners that "the revolutionary government has taken immediate and concrete measures to rehabilitate thousands of people," without specifying just what those measures were. Just as widespread failure of the main harvest following the June–September rains was pushing disaster into catastrophe, the Ethiopian government turned its back on its starving subjects. It directed all attention and resources toward a lavish celebration of the tenth anniversary of the revolution that had brought it to power. It followed this up with launching the political party for which Soviet and Soviet-allied countries had been pushing—the Workers' Party of Ethiopia (WPE). It prevented Western journalists from traveling north to cover the famine.[57] For their part, Western governments and aid organizations had been slow to mobilize food relief. There was continuing skepticism about the accuracy of RRC estimates of at-risk populations and needed relief. The RRC was suspected of inflating figures, and behind it stood the Ethiopian government, in which potential donors had little confidence. They doubted the RRC's capacity to manage and deliver aid and tried to negotiate numbers that might be palatable to their constituents. The RRC's estimates turned out to be more reliable than relief agencies initially thought.

At the beginning of October, the politburo of the newly founded WPE faced the situation and adopted famine relief as a major campaign. It gave permission for foreign journalists to visit Wallo and Tigray; toward the end of the month, British TV coverage galvanized world public opinion, which in turn forced Western and international institutions to act. Within two weeks, at the beginning of November, "20 fixed-wing aircraft and 30 helicopters from UK, USA, USSR, FRG, GDR, Italy and Libya" were airlifting supplies from Asmara, Addis Ababa, and Assab to central distribution points in Tigray.[58] NGOs, a number of which had already been active, stepped to the fore and, under the supervision of the RRC, itself now closely monitored by the politburo, became the principal conduits for international relief. Table 2.2 lists the most prominent and figure 2.1 shows

TABLE 2.2. NGOs active, 1984–1985

CARE
CDAA, a pool of US NGOs
CONCERN, Irish NGO
CRDA, Christian Relief and Development Association, a pool for smaller NGOs
CRS, Catholic Relief Services
ECS, Ethiopian Catholic Secretariat
EOC, Ethiopian Orthodox Church
ICRC, International Committee of the Red Cross
IMT, Italian Medical Team
LRCS, League of the Red Cross Societies
LWF, Lutheran World Federation
MSF, Médicens sans Frontières (French and Belgian)
Oxfam, English NGO
SCF, Save the Children Fund (English and US)
WUSC, World University Service of Canada
WV, World Vision International (US)

Source: Jansson, Harris, and Penrose, *The Ethiopian Famine*, xiii.

Figure 2.1. Major NGOs and main areas of activities
Source: Jansson, Harris, and Penrose, *The Ethiopian Famine*, xiii.

their geographical distribution. Through a whirlwind of recriminations and accusations, international relief was forthcoming.

The famine was truly of major proportions. One sober estimate suggests that 500,000 people died as a result; another reduces the total to 400,000, of whom 100,000 were in Wallo.[59] We ought not to expect greater precision, but it should be clear that the famine of 1984–85 was much more devastating than its predecessor in 1973–74. Although many died—some from starvation, some from conditions brought on by malnutrition, others from exposure, and still others from epidemic diseases—many more victims were reduced to the collapse of their livelihood and the prospect of starvation.[60] Table 2.3 presents the official RRC numbers of famine victims. In 1985 they numbered roughly 8,000,000 people and in 1986 still amounted to 5,915,000 people. As in 1974 the most seriously affected provinces were Wallo, Tigray, Eritrea, and Hararghe. These provinces accounted for about three-quarters of all famine victims in 1985 and 1986 (see table 2.4). In total numbers, Wallo suffered most in 1985 and 1986, with 2,590,000 victims, 71.7 percent of its population in 1985. The Wallo numbers for 1986 are 1,455,000 and 40.3 percent. Famine was also serious in Tigray. In 1986 the percentage of its population still starving was higher than that of any other province. Famine did not strike uniformly. Both times it hit first and hardest in the *qolla* areas in the east. In the highlands, too, shortages of rainfall led to famine occurred arbitrarily.[61]

Given the contentious national and international politics that surrounded the famine, the numbers were controversial. Large parts of Eritrea and Tigray were outside government control, so although there is greater uncertainty about their reported numbers, their being active war zones suggests more rather than fewer victims.[62] By contrast the Wallo numbers gain some credence since much of the province was under government control. All sides agreed that Darg control throughout the country was stronger and more direct than had been the control of Haile Selassie's government. It proved able to conduct the country's first national census in May 1984, the results of which were as persuasive as they were unexpected. The census reckoned the country's population to be about 42 million, considerably larger than previously thought.[63] This would make the 8 million "drought-affected" people in 1985 to be 20 percent of the total population. Not that Darg numbers were free of political manipulation, but there is a consistency in the growth of the numbers of famine victims from 1983 onward that gives them some utility.

TABLE 2.3. Drought-affected populations, 1985–1986

	Population	*Drought-affected people*			
	1984	*1985*		*1986*	
Province		*Number of drought-affected people*	*% of total provincial population*	*Number of drought-affected people*	*% of total provincial population*
Eritrea	2,615,000	830,000	31.7	850,000	32.5
Gondar	2,905,000	365,000	12.6	135,000	4.6
Tigray	2,410,000	1,430,000	59.3	1,000,000	41.5
Wallo	3,610,000	2,590,000	71.7	1,455,000	40.3
Shawa	8,090,000	850,000	10.5	630,000	7.8
Sidamo	3,790,000	535,000	14.1	445,000	11.7
Hararghe	4,150,000	875,000	21.1	1,200,000	28.9
Subtotal	27,470,000	7,475,000	27.2	5,715,000	20.8
Others		535,000		400,000	
Total		8,000,000		5,915,000	

Source: Jansson, Harris, and Penrose, *The Ethiopian Famine*, xii. Population numbers are from the Ethiopian Relief and Rehabilitation Commission and are approximate.

TABLE 2.4. Most seriously affected provinces, 1985–1986

Province	*1985*		*1986*	
	Total population	*% ot total population affected by famine*	*Total population*	*% of total population affected by famine*
Wallo	2,590,000	71.7	1,455,000	40.3
Tigray	1,430,000	59.3	1,000,000	41.5
Eritrea	830,000	31.7	850,000	32.5
Hararghe	875,000	21.1	1,200,000	28.9
Total	5,225,000		4,505,000	
% of national total famine victims	71.6		76.1	

Source: Jansson, Harris, and Penrose, *The Ethiopian Famine*, xii.

Relief flowed in at high levels through 1985 and 1986. Drawing on reports from the principal UN agencies involved in relief, Dessalegn estimates that $906,070,000 worth of emergency food aid flowed into Ethiopia in 1984–85 and 1986. Of this total, two-thirds came by the end of 1985 and the remaining third the following year. Table 2.5 presents the information.[64] Wallo received a generous proportion of aid. Throughout 1985 the RRC received about 24,000 metric tons of grain a month. Of this some 9,000 metric tons were distributed in Wallo. In Ethiopian terms the influx of resources was prodigious. Depending on the year and exchange rate used, it equaled anywhere from 16.9 percent of Ethiopia's national government expenditures to 89.8 percent. Hansson reports central government expenditures in 1984–45 as birr 3,640,000,000 and in 1985–86 as birr 4,065,000,000. At the official exchange rate of B2.07 = US$1, this would make foreign food aid equal to 33.8 percent of total Ethiopian government expenditures in 1984–85, and 16.9 percent in 1986. However, in 1984–86 the birr was overvalued. The first open market exposure of the birr, which took place in 1993 and 1994, pegged it between B5–6 = US$1. Averaging this rate at B5.5 and projecting it back to 1984–86 would make the value of foreign food aid equal to 89.8 percent of total Ethiopian government expenditures in 1984–85 and 42.2 percent in 1986.

TABLE 2.5. Emergency food aid to Ethiopia, 1984/85 to 1986 (in US dollars)

Donors	*1984/85*	*1986*	*Total*
United States	252,069	126,000	378,069
European Economic Community*	59,951	72,800	132,751
Canada	41,974	20,200	62,174
Soviet Union	52,000	—	52,000
France	23,045	3,200	26,245
West Germany	17,124	9,059	26,183
Australia	14,474	6,082	20,556
India	15,390	1,600	16,990
Other countries	89,746	51,146	140,892
UN organizations	28,354	21,856	50,210
Total	594,127	311,943	906,070

Source: Adapted from Dessalegn, *Famine and Survival Strategies*, Annex 6 (246–47), with corrections.

* In addition to collective EEC relief and the contributions of France and West Germany, EEC countries contributing relief were Belgium, Denmark, Greece, Ireland, Italy, the Netherlands, and the United Kingdom.

In a broadcast on February 9, 1985, Mengistu Haile Mariam outlined an Ethiopian response to the crisis. The government introduced a special famine relief tax, levying one month's pay from every Ethiopian. For farmers, famine-afflicted or not, this was pegged at 100 kg of grain. Luxury imports, mostly textiles and cars, were banned, and strict gasoline rationing was introduced. A centerpiece of the government's response to famine was the resettlement of farmers from Tigray and Wallo to more lightly populated areas in western and southwestern Ethiopia, the principal "immediate and concrete measure" to which he had referred in his speech the previous June. Mengistu enjoined all Ethiopians to be prepared to work to set up these settlements. Before the end of the school year he suspended university instruction and sent faculty, staff, and students to the resettlement areas.[65]

Resettlement was not a new program. In its last years, Haile Selassie's government resettled some 10,000 families. Most of these settlements failed, and their participants abandoned them. In a continuation of the program, during the first ten years under the Darg, some 187,000 people were resettled. Like the schemes of the imperial government, these were generally judged a failure, because of poor planning, bad site selection, and the imposition of collectivization as the preferred organization for settlers. In October 1984 the Darg returned to resettlement as its principal response to famine, without having learned any of the lessons of earlier failures. From October onward, the Darg relentlessly pursued the goal of resettling 1.5 million farmers, drawn principally from Tigray and Wallo. It redirected the RRC and all its resources away from food relief in the famine-stricken north to running the resettlement program. Most farmers were resettled along the western borderlands in Wallaga and Illubabor, where population densities were much lighter. By January 1986, when the program was suspended, it had resettled some 500,000 farmers.[66]

While the Darg was refocusing relief efforts toward resettlement, food aid began to arrive, and mortality in the camps began to decline. In Korem, perhaps the most intensively reported relief camp, deaths were about 100 a day in October 1984. By mid-1985, "only a few were dying each day." In 1986 the situation was much improved. Although food aid remained at a high level and the population in the relief camps remained high, the camps were starting to empty out because of resettlement and the increasing distribution of relief to people in their home locations.[67] By November 1986, the US press noted the slow passing of the famine. Ian Steele reported to the *Baltimore Sun*, under the headline "Ethiopia expects crops to be best in five years," that the country would be "off Africa's critical list by Christmas. The refugee camps have closed, and food lines, which absorbed close to $1 billion worth of grain and other international assistance in 1985, are now the exception in most parts of the country." A food deficit of about 2.4 million tons in 1985 and 1.2 million tons in 1986 had now been lowered to a projected 400,000 in 1987. While the United States was precluded by federal law from participating in any long-term development projects, other Western countries and the World Bank were offering continued relief conditional on the Ethiopian government liberalizing its agricultural policy "to establish more flexible pricing arrangements for its farmers and to allow the free movement of agricultural produce between areas of surplus and deficit."[68]

Aftermath

The story did not end there, nor has it ended yet. The famine of 1984–85, like that of 1973–74, was followed by a revolutionary transformation of the Ethiopian state. In 1991 the Darg was violently overthrown by the EPRDF, a creation of the Tigray People's Liberation Front. The EPRDF dismantled the state structures created by the Darg and replaced them, in 1995, with a newly constituted Federal Democratic Republic of Ethiopia (FDRE).[69] It rejected a host of Darg policies but accepted land reform, much as the Darg had instituted it, and the Darg's analysis of the relationship between famine vulnerability and environmental stress.

The rains in the late 1980s and for much of the 1990s supported increases in crops. But famine vulnerability did not disappear. In 1990–92 drought and warfare apparently threatened 4 million people in northern Ethiopia with serious food shortage. In 1993–94 new droughts contributed to the need for increased food assistance in Tigray and Wallo. In 1999–2000, drought once again reduced crop yields and more than 100,000 people received relief aid. In 2002 "erratic and inadequate rainfall patterns" reduced crop production by 21 percent below the preceding five-year average, increased livestock mortality, and threatened more than 14 million people with a return of famine. Food assistance was called for—1.5 million tons. In 2003 resettlement was revived as a grand strategy to eliminate famine vulnerability.[70] This strategy rested on a narrative of the environmental degradation of the northern Ethiopian highlands as the product of destructive farming practices brought about by overpopulation. Chapter 3 explores this narrative and the interventions it gave rise to.

3

Environment and Famine

Explanations and Interventions

> The highlands of Ethiopia . . . contain what is probably one of the largest areas of ecological degradation in Africa, if not in the world. . . . It's already a moonscape in the central highlands.
>
> —*Los Angeles Times*, May 11, 1984

So wrote an anonymous expert articulating an apocalyptic vision of the Ethiopian highlands in a report issued by the UN Food and Agricultural Organization (FAO)—overheated, to be sure, but representative of a good deal of expert perception. Wallo is paradigmatic of the famine-vulnerable highlands, yet some ten years later colleagues of mine, visiting Wallo for the first time, were impressed by how verdant it was—crops flourishing and ripening, brooks flowing. One described it as a "Garden of Eden."[1] Circumstances profoundly shape perception. The anonymous FAO expert reflects a period of drought and the knowledge of incipient famine. My colleagues reflect a moment of recovery in a landscape toward the end of an ample rainy season. Presumably the "real" state of the highlands lay somewhere between these extremes, but the apocalyptic view gained the support of Ethiopia's governing elite and international experts and led directly to a host of landscape interventions and the dislocation of some 500,000 people from Wallo and Tigray to the Sudan borderlands (see chapter 2).

The apocalyptic view prevailed throughout the 1980s and grounded explanations of famine vulnerability and the landscape interventions inspired by these explanations, the value of which was held to be self-evident. The inference from the apocalyptic landscape was that the principal sources of landscape "degradation," and hence famine vulnerability, were "traditional" farming practices and population growth.[2] Some observers thought that warfare and "socialist" economic practice increased famine vulnerability.

Farming Practice

In the 1980s, two internationally backed projects profoundly shaped official thinking in Ethiopia about the nature of environmental change and the role of farmers in bringing it about. In doing so, they drew on global narratives. When famine broke out in 1984, they were in a position to provide a complete rationale of what was going on and a plan of action for dealing with it.

The Soil Conservation Research Project (SCRP) started work in 1981 under the direction of Hans Hurni, a Swiss geographer. The SCRP had the financial support of the Swiss government. Its approach was national in scope, and it set up research stations sited to represent six major agro-climatic zones. The stations were to carry out controlled experiments to document erosion processes. One of the stations was in Wallo near Dessie.[3] In Hurni's eyes, the situation was truly dire.[4] Erosion stripped 1.5 billion tons of soil a year from the Ethiopian landscape. Within 100 to 150 years, all of Ethiopia's topsoil would be gone.

In 1983, two years after the founding of the SCRP, the Ethiopian Highlands Reclamation Study (EHRS), a joint project of the FAO and the Ethiopian Ministry of Agriculture with funding from the World Bank, was launched. It entailed 300 person-months of expertise in fifteen subjects over a two-year period from mid-1983 to 1985. Within a year it boldly predicted that "18 percent of the highlands will be bare rock by 2010 and 10 million people will not be able to produce food from the land," a prediction that found its way into an account of the study's major findings and recommendations, which also warned, "For today's children over a third of the highlands could become incapable of sustaining cropping."[5] These dire predictions reflected the preconceptions of the study creators, who had already determined that the highlands needed to be "reclaimed."[6]

The recommendations of the EHRS, at least as presented by Constable and Belshaw, were innocuous enough. Conservation activities ought to be "integrated into mainstream agricultural and rural development activities" (144). It recommended a conservation-based development strategy (CDS) to operate at four levels: (1) Farmer land-use practices should promote increased vegetative cover of the soil; (2) the national agricultural development strategy should support the "core component" of the CDS; (3) a multisectoral approach to rural areas should provide infrastructure, promote voluntary resettlement, provide for popular participation in formulating and implementing conservation plans, and promote employment diversification; and (4) national development planning and policy formulation should direct adequate resources to the CDS.[7]

The EHRS recognized three major agro-ecological zones according to criteria that privileged production: high-potential perennial crops; high-potential cereal crops; and low-potential cereal crops. These bore little resemblance to the culturally sanctioned landscape categories of *dagga*, *wayna dagga*, and *dagga*. The EHRS did incorporate altitude in its further refinement of its major agro-ecological zones into nine subzones. Both the SCRP and EHRS defined the problem as the "degradation of resources," which was further defined as "reduction in their longer term productivity" and argued that it was caused primarily by soil erosion. At the time of the study, in findings that echoed those of the SCRP, the EHRS reckoned fully one-half of the highlands to be "significantly eroded, a further quarter to be "seriously eroded," and 3.7 percent of farmland to be "so badly eroded that they were unlikely to sustain cropping in the future."[8] Soil erosion, in turn, was the product of the "removal of natural vegetation for cropping, grazing, fuel and building," and the cause of vegetation removal was increasing population, which expanded the area under cultivation to land unsuitable for cultivation, mainly because of the steepness of slope.[9] This diagnosis was Malthusian—population increase had driven farmers to the point that they were now mining the environment, extracting resources at a rate that far outstripped the environment's capacity to recover. Population exceeded the land's "carrying capacity."

Both government officials and international experts agreed that the highlands suffered from a degradation of resources. Their diagnoses of the underlying problem were convergent to the extent of agreeing that changing Ethiopian farming practices was required. The international experts saw the problem as caused by waterborne soil erosion, of which four-fifths was produced by "overexploitation" of cropland. Most of the rest was caused by rural Ethiopians, who were responsible for "overgrazed grasslands, wastelands and newly deforested areas."[10] The government was more concerned with "backward" farming practices in general. From its standpoint, modern science and technology under the guidance of scientific socialism and new forms of production—state farms and producers' cooperatives—

would bring the Ethiopian countryside into the twentieth century and guarantee the country's future prosperity.

Analyses of the SCRP and the EHRS have met with substantial criticism. Specifically, later commentators argued that the two projects much exaggerated the extent of soil erosion.[11] More generally, their diagnoses rested on an inadequate grasp of the history of Ethiopian landscape change.[12] While they gestured to the need to be attentive to existing farmer practices and involve farmers in the intervention design, they proved unable or unwilling to do either. They overlooked or undervalued the conservation practices already used by farmers, who maintained grass bunds, built drainage ditches on the contour, terraced, used fertilizers, and managed common meadow lands by local customs and institutions.[13] These practices were not always optimally deployed, and some existing practices—mono-cropping, multiple plowing of the land in preparation for sowing cereals, bare fields at the time of heaviest and most erosive rainfall, not to mention cultivating steep slopes—did contribute to erosion, but not to the extent depicted in the influential accounts by the international experts.

While the government was essentially interested in the revolutionary transformation of Ethiopian agriculture, in the interim it found common cause with the international experts in focusing on environmental interventions designed to stop and combat erosion. To deal with it, in consultation with international experts the Ethiopian government developed "a massive programme for afforestation and soil conservation," which pursued four lines of landscape reclamation: terracing, or bunding, of farmland; afforestation; hillside enclosure; and gully control.[14] Contrary to the EHRS recommendation, at least in Wallo, this was not folded into a grand plan of national agricultural development. In its development plans, the government remained resolutely focused on what the EHRS had designated as areas of high agricultural potential and, within them, on state farms and agricultural producer cooperatives.

Some afforestation and soil conservation practices dated to the 1960s and early '70s and had increased under the Darg in the later '70s. Following the outbreak of famine, they expanded rapidly. Local peasant association personnel along with agents of the Ministry of Agriculture, organized through the Community Forests and Soil Conservation Development Department, cajoled and induced farmers into participating. The World Food Program provided the inducement with its Food for Work program, which paid participants in wheat and vegetable oil. The scale was prodigious. Ethiopian farmers played no role in setting the goals or the design of the conservation projects.

The program had first been formalized in 1980 as the World Food Program, "Project Ethiopia 2488—Rehabilitation of Forest, Grazing and Agricultural Lands." Between 1976 and 1985 it created 470,000 kilometers of hillside terraces for afforestation, thousands of check-dams in gullies, thousands of kilometers of rural roads, and almost 80,000 hectares of areas closed for natural regeneration. By 1990 it claimed over 1 million kilometers of bunding and 300,000 hectares reforested. More than 800,000 farmers, men and women, provided the labor. By 1992 the project had been supported by more than 720,000 metric tons of wheat and 25,000 metric tons of vegetable oil. The total value of food donated through the World Food Program between 1975 and 1990 was just over US$250 million.[15] Overall, Dessalegn estimates that US$900 million was invested in environmental protection between 1974 and 1993. Donors in descending order of contribution were the government of Ethiopia (US$300 million), the World Food Program (US$275 million), bilateral donors (US$100 million), UN agencies (US$50 million), and NGOs (US$25 million).[16] The landscape impact of afforestation and terracing—large eucalyptus plantations, enclosed hillsides, and terraces everywhere—was evident to anyone traveling in rural Ethiopia in the mid to late 1980s. It was not limited to arterial roads but was evident along seasonal and all-weather roads.

Afforestation gave rise to two different kinds of forests—one under direct state control, the other under the ostensible control of local communities. While existing woodlands were taken over, new state forests were established, most notably, to a casual observer, on the enclosed hillsides, and less obviously on farmland and pasture. The state forests were completely closed to local farmers. The community forests were supposed to be controlled by local peasant associations, whose members would benefit from less erosion and controlled access to the resources of the forest itself. This did not happen. Farmers had little or no access to the community forests.[17]

What Ethiopian farmers thought of government-imposed conservation was revealed in 1989 and 1990, following the Tigray People's Liberation Front (TPLF) occupation of parts of Wallo and the government's announcement of a change in economic policy away from a command to a mixed economy. The new economic policy encouraged the private sector.[18] It increased the security of landholding, explicitly allowing the transfer of land to legal heirs and recognizing the ownership rights of farmers over trees. It further recognized the rights of farmers to hire laborers and dissolve producers' cooperatives. It eliminated the quota system and controls on the grain trade and allowed private traders to compete on an equal footing with state agencies. This announcement, together with a rapid erosion in the government's military position vis-à-vis the TPLF, undermined its authority at the local level. The result was an assault on the state-sponsored afforestation and erosion control schemes, which started in anticipation of the arrival of TPLF troops and was boosted by the change of economic policy and continued through to the establishment of the new government in 1991.

From northern Shawa to northern Wallo, farmers plowed through the bunds, which the authorities had required them to construct on their land. They plundered the afforested hillside plantations, hauling off the lumber for their own uses and leaving some of the plantations beyond recovery. A notable victim of this onslaught was the "One Million Tree" plantation, a showpiece of "participatory" conservation, sponsored by the Upper Mille and Cheleka Catchments Disaster Prevention Program. One informant noted that the trees had been uprooted "only after foreign dignitaries began arriving on visits" and went on to say that this drew attention to the area and "the decision was made to have more plantations." Deforestation flourished in the space and time between the withdrawal of the Ethiopian national army and the arrival of the TPLF forces. Although fighting and looting by soldiers destroyed some of the plantations, local farmers inflicted most of the damage. As one of them acknowledged, "The massive destruction of the forests was carried out by the farmers themselves."[19]

Farmers plowed up the bunds in their fields, because, in their experience, bunding was counterproductive and a waste of their time and labor. The conservation recommendations had been developed by the Soil Conservation Research Project on the basis of controlled experiments. The results did not translate easily to farmer practice. In too many cases, the measures lowered productivity. They took too much land out of production, encouraged water-logging; harbored rats, which ate ripening crops; and created fields too narrow for efficient plowing.[20] Farmers were not hostile to trees as such. Privately planted trees emerged unscathed. Destruction of government-sponsored conservation interventions was an expression of the rage farmers felt at the government's appropriation of assets entailed by enclosure and afforestation. Farmers could no longer graze cattle on the hillsides too steep to farm, nor could they gather grasses for thatching, fodder, or other household uses. They lost access to fuel and medicinal plants.

The government's long-term solution for what it continued to view as the overpopulated, chronically drought-prone and environmentally degraded farmlands of Tigray and Wallo was the resettlement of farmers from Tigray and Wallo to less densely populated parts of the country in the west and southwest—Gojjam, Wallaga, and Illubabor—areas it deemed underpopulated, environmentally robust, and of high agricultural potential. The western lands were malarial, already under occupation by the Oromo and a variety of Omotic and Nilotic peoples, and poorly surveyed, but these were not impediments for the government.

Almost from the beginning, resettlement came in for heavy criticism. It was held to be coercive. This was undoubtedly the case. Moreover, since recruitment took place at feeding centers and the peasant associations, many of the resettled would have been in no position to make a choice. Clay and Holcomb interviewed 277 Ethiopians who had been resettled and subsequently escaped as refugees to the Sudan: "All those interviewed insisted that they had been captured by government troops and forced to re-settle and that considerable coercion was used by the government to force individuals to take part in the program." Alula Pankhurst gives a very different impression on the basis of interviews with people who were still in the settlements to which they had been relocated, but he also cites a government survey as finding 38 percent of the resettled as having been coerced and gives additional reasons as to why this figure could well have been higher.[21] Moreover, the real objective of resettlement, critics alleged, was to undermine the oppositional TPLF by depopulating the Tigray highlands. Although this should not be discounted as one calculation among many the authorities made, it would appear that a clear majority of the resettled came from Wallo, beyond the zone of TPLF conflict.[22] Nor was it easy for the government to recruit and coerce Tigray farmers given how much of the Tigray countryside was controlled by the TPLF.

Fundamental to the resettlement project was the environmental perception that the agricultural potential of the highlands of northern Shawa, Wallo, and Tigray had been "degraded" to the point of destruction by backward farming practices. The abuses of the program notwithstanding, in 1986 the Ethiopian Highlands Reclamation Study continued to endorse the strategy of resettlement. "Resettlement is necessary in order to provide breathing space for reclamation and even conservation to be effective in the most densely populated and degraded areas. . . . It is estimated that considerably more than 150,000 persons annually will need to be resettled, or preferably 'migrate' voluntarily if present population growth and degradation trends continue."[23] The degradation narrative, which provided the underpinning for this program of environmental transformation, was not wholly divorced from reality. Inherited agricultural practice had proved unable to respond to the challenge of severe drought. The Ethiopian agricultural landscape was indeed under stress and not just from drought. What was missing from the narrative was any sense that Ethiopian farmers might already have developed effective coping strategies or that they had anything to contribute. What was equally missing was any sense that externally imposed conservation measures might be counterproductive. Science would solve the problem. Experts would show the way.

Drought

Chapter 1 argued that the principal environmental trigger for the Great Famine of 1888–92 was microbial—a virgin rinderpest epidemic. The principal environmental factor underlying the famines of 1973–74 and 1984–85 was drought—acute shortfalls of rain during the growing seasons. Balancing the relative importance of drought vis-à-vis other factors is a fraught but still productive exercise.

Some parts of Ethiopia are more drought-prone than others, and the most drought-prone parts of the country are also the most famine-prone—the northern provinces of Eritrea, Tigray, Wallo, and northern Shawa; and eastern Ethiopia, the Ogaden, and neighboring provinces. Drought, as the major precipitant of famine in 1984, is a constant theme in contemporary news reports and retrospective accounts by relief workers. According to the informants of Alemneh Dejene, who carried out field research in Wallo from October 1987 to April 1988, "shortage of rainfall" was "the primary reason for crop failure leading to famine" in 1984. The droughts of 1973 and 1984 affected all parts of Ethiopia, although they did not bring famine everywhere.[24]

In the past two decades, meteorological studies of Ethiopia have proliferated attempting to make sense of the possible relationships between drought and famine from about 1965 onward, the point in time at which higher quality climate data become available.[25] These studies look for patterns of climate change in the recent past and attempt to anticipate change in the twenty-first century. They take as their point of departure the fact that drought brought on by "negative rainfall anomalies" is a major contributor to famine vulnerability in Ethiopia. At the same time, noting how Africa in general and Ethiopia in particular are often held to be extreme examples of what we may expect from "anthropogenic climate change," these studies recognize that drought or otherwise induced crop failure alone cannot account for such an extreme outcome; many interacting factors contribute to the complex nature of vulnerability to climate shocks.[26]

Although the total amount of rainfall in any given year is important, equally important is the seasonal distribution of rain in historically rooted patterns. This allows for a distinction between meteorological drought and agricultural drought.[27] The former is an absolute shortfall in precipitation measured against a longer-term mean, the latter a shortfall in the precipitation of seasonal importance for agriculture. Agricultural drought does not require meteorological drought, although they often go together. We have seen that there are two seasons of rain of agricultural importance in our part of Ethiopia—the *balg*, or lesser rains, which start in February and may continue to May; and the *keremt*, the major rains, which set in in June and continue through into September. Throughout the area of study in this volume, *keremt* is of overwhelming importance, feeding the crops that are harvested in late October through November. In some parts of Ethiopia, the *keremt* rains contribute up to 95 percent of total annual precipitation and nationally may feed the production of anywhere from 85 to 100 percent of Ethiopia's annual crop.[28] However, in many if not all of the heavily settled *wayna dagga* regions in Wallo, the *balg* crop is one on which farmers have become dependent, despite its apparently modest contribution to their total production. *Balg* rainfall in these regions is above the national average. In the central Wallo towns of Kombolcha, Dessie, and Kamise, the *balg* rains contributed on average a little over 20 percent of the annual total rainfall during the years 1975 to 2003.[29]

The general dynamics of the atmospheric systems, which shape Ethiopia's climate, are fairly well established. The *keremt* rains occur during the North American summer and are driven by the northward migration of the Intertropical Convergence Zone (ITCZ), which draws moisture-laden air from the tropical Atlantic northward over West Africa and across the Sudanic belt as far east as the Ethiopian highlands, where it rises in response to the upward lift in topography and precipitates rain. As figure 3.1 illustrates, this system brings rainfall that is heaviest in southwestern Ethiopia and declines steadily as it moves to the northeast and the eastern edge of the great plateau. Here it encounters a landscape dropping dramatically away to the east. It encounters climate dynamics associated with the Indian Ocean monsoons and the El Niño–Southern Oscillation of the Pacific Ocean. Because of its association with the ITCZ, the timing and amount of rainfall over the Ethiopian highlands during the North American summer sometimes parallels that of the African Sudanic belt, which runs westward across the continent to Senegal. This was the case in 1972 and 1984, when drought affected the Sahel and the Ethiopian highlands. In the 1990s, however, rainfall patterns diverged with dryness continuing to afflict the Sahel, whereas Ethiopia returned to the higher levels of rainfall it had enjoyed in the 1950s and earlier.[30] Variability in this system derives from its timing, the pace of its advance and retreat, and the northward distance that it reaches.

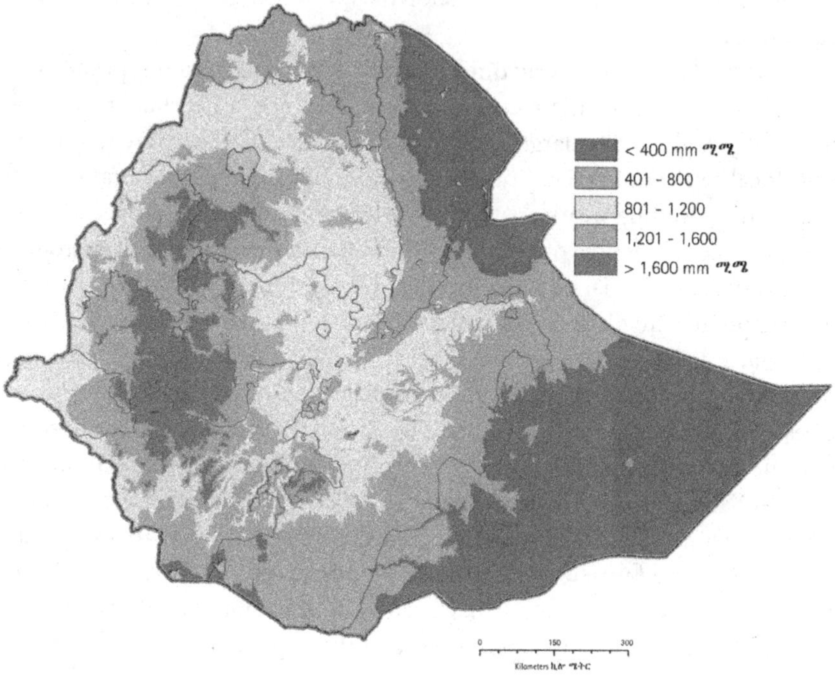

Figure 3.1. Annual distribution of precipitation in Ethiopia
Source: Tadesse, Alemu, Bekele, Tebikew, Chamberlin, and Benson, *Atlas of the Ethiopian Rural Economy*, 19. Reproduced with permission from the International Food Policy Research Institute, www.ifpri.org. The original figure is available online at https://www.ifpri.org/publication/atlas-ethiopian-rural-economy.

The moisture-laden air, originating over the Indian Ocean, which waters southern and southeastern Ethiopia, also brings *balg* rain to Wallo and other easternmost parts of the highland plateau. Ultimately, these great climate systems are correlated to sea surface temperatures in the Atlantic, Pacific, and Indian Oceans. There appear to be no statistically significant trends toward overall increases or decreases in total rainfall in Ethiopia over 1965 to 2002, although there may be a local increase in annual and *keremt* rainfall over 1961–2003 at Dessie. There is consensus on the variability of rainfall on a year-to-year and decade-to-decade basis. Conway and Schipper reckon the range of variability to be about 45 percent, from –26 percent to +19 percent.[31] The 1980s were a drier decade than the long-term mean, whereas the 1990s were wetter. Climatologists appear to agree that there is a trend toward rising temperature. If maintained, ultimately this will have consequences for agriculture, but it would appear that this is not yet the case. Since the late 1950s, shortfalls of rain have been the key precipitant of famine.

Woldeamlak and Conway draw attention to spatial and temporal variability. They note that "high levels of spatial variability exist that are unlikely to be fully explained by large scale climate influences." There are many possible local sources of variability—topography, altitude, and orientation with respect to prevailing weather systems not least among them. Surface cover—the presence or absence of vegetation, the nature of vegetation if present—is yet another potential factor, one favored by Alemneh.[32] As "deforestation," it is frequently invoked in discussions of Ethiopian famine vulnerability.

Surface cover change is at the heart of a powerful narrative that sees drought as a result of "desertification" and/or "deforestation." Cutting down trees and removing other forms of vegetation by pastoralists and farmers became a major explanation of the drought afflicting Ethiopia and the African Sahel in 1973. A UN Conference on Desertification in 1977 blamed environmental degradation, leading to desertification, on "overexploitation" of land by farmers, herders, and woodcutters.[33] In Africa, according to the narrative, this process had led to the inexorable expansion of the Sahara. In the case of Ethiopia, it became the explanation favored by politicians, international consultants, and aid workers to account for drought and famine in the early 1970s and its grip on the official mind was only intensified by the return of famine in the 1980s.

The belief in a link between deforestation and drought dates to antiquity but owes its modern origins to the eighteenth century and the creation of

European tropical empires. Naturalists inferred from the deforestation of tropical islands like Barbados and Mauritius that drought was the result of the removal of natural vegetation.[34] This explanation in turn influenced the policies of some colonial governments and resulted in the creation of forest preserves and an interest in tree planting. Between the eighteenth and twentieth centuries, the link between deforestation, devegetation, and drought was scaled up from the level of an island to that of a subcontinent. It became an animating anxiety for colonial officials in French West Africa in the 1930s and was picked up by some of their British colleagues, retaining sufficient heft to be brought back into play to explain drought-induced famine in the 1970s in the Sahel and by extension Ethiopia.[35] Its role was overplayed. Rather than "inexorably" expanding, another view of the Sahara is that it could just as well be perceived as oscillating with vegetation recovery following a return to more "normal" patterns of rainfall.[36] Figure 3.2 presents the changes of precipitation over northern Ethiopia during the thirty-year period from 1953 to 1983. It is difficult to see how changes in surface vegetation, particularly over a year-to-year and decade-to-decade timeframe, could adequately explain this pattern. Nevertheless, changes in

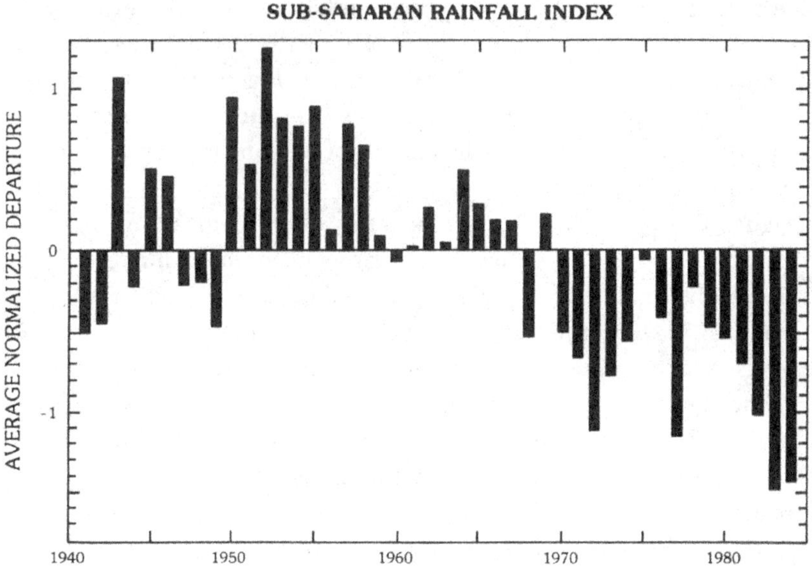

Figure 3.2. Northern Ethiopia annual rainfall departures, 1953–1984
Source: Degefu, "Some Aspects of Meteorological Drought in Ethiopia," 33.

surface vegetation may well account for some of the local and regional variation in Ethiopian rainfall patterns. Chapter 4 assesses the evidence for vegetation change in Wallo.

Workineh Degefu provides a standard account of the climate events that led to famine in northeastern Ethiopia in 1984.[37] He starts in 1982 when the onset of the *keremt* rains was delayed by two months, producing crop failure in Gondar and Wallo provinces. He describes the rains in 1983 as "generally good" in both growing seasons, if slightly below normal in Wallo. In general October 1983 would appear to have been unusually and unseasonably wet. A prolonged dry spell then set in and brought the failure of most of the *balg* rain—in February, March, and April 1984. That season ended with heavy, but untimely rain in May, which precluded normal planting. The dry season, which normally characterized late May and early June, was replaced by the onset of the *keremt* rains, which continued into June and July, "with decreased amounts and with poor spatial and temporal distribution," leading to a dry spell in late July, the point at which the *keremt* rains normally reached their climax. Drought continued into August and September, only to be alleviated by untimely heavy rain in October, which in some places continued into November, when harvesting should have been getting under way. Not only was rainfall unseasonal, but the overall amount of rain falling in 1984 was far below the twenty-one-year mean. As we will see, the pattern laid out by Workineh is generally reflected in the detailed climate information, which we have for a number of sites in Wallo. About the severity of drought in 1984 or of its contribution to famine in that year there is no controversy.

Figure 3.2 shows that the famine years of 1965, 1973, and 1984 were marked by sharp declines from recent norms in annual rainfall. The extent to which the severity of any given year's falloff in rain and its impact on agriculture may be intensified by declining rainfall in preceding years remains an open question. The rainfall deficit in 1984 was greater than in 1973 and the famine of 1984 was greater in the area and numbers of people it affected. Was this proportionate to the intensity of the annual fall-off? This is unclear. Overall, Conway notes, the droughts of the 1970s and 1980s were no worse than individual years in the longer record of Ethiopian climate, but they "were unprecedented in terms of the prolonged period of successively drier than average years."[38] The figure indicates that 1973 was the third in a sequence of increasingly dry years, which may have intensified its impact. The figure also suggests that 1984 was preceded from 1978 onward (except for 1982), by five years of rainfall below the long-term norm, although the extent of the deviation in these years never approached the intensity of the fall-off in 1971 and 1972. The figure is based on national data, and as we have seen, there is considerable local variation because of the complexity with which the large-scale climate systems interact with each other and with Ethiopia's rugged topography.

Wallo, however, provides some good illustrations of how local realities may reflect or defy national trends. Altogether we have climate information for six sites in Wallo: Dessie, the regional capital; Kombolcha, the principal airport in the region; Bati, a market town to the east; Rike and Kamise, towns 66 and 75 kilometers to the south of Dessie; and Qobo, another market town, about 170 kilometers north of Dessie.[39] They are of particular interest to Tahuladare, that part of Wallo on which chapters 4 and 5 concentrate, and they fall within a larger cluster of sites whose climates show similarities in terms of the amount of annual rainfall they receive. This cluster, which generally lies along the eastern edge of the Ethiopian plateau, is distinguished from another cluster in the Amhara regional state to the west, around Lake Tana.

The climate records cover different time periods—1961–2003 for Dessie, Kombolcha, and Kamise; 1966–88 for Bati and Rike; and 1977–88 for Qobo—which yield somewhat different annual means. Nonetheless, they allow for some comparisons. For example, five of these sites experienced major shortfalls of rain in 1984, particularly in the principal agricultural seasons of *balg* and *keremt*. During that year Bati received only 37.4 percent of the annual mean rainfall over 1968–88, Rike received 52.3 percent, and Qobo received 19.1 percent. (For the deviations from the 1968–88 mean in Bati, Rike, and Qobo, see table 3.1.) In that same year, annual rainfall in Kombolcha and Dessie "was 2.6 and 2.7 times the standard deviation below the 1961–2003 mean."[40] Remarkably, Kamise rainfall in 1984 appears to have been slightly higher than its annual mean from 1961 to 2003.

The graphs in figures 3.3 and 3.4, which show the patterns of rainfall in Bati and Rike throughout the year 1984, are examples of agricultural drought associated with meteorological drought. They show a similar pattern of annual distribution of rainfall, except for December. In Bati the total rainfall for 1984 was 37 percent of the long-term mean annual rainfall. The *balg* rainfall in February through April ranged from 10.6 percent of the long-term mean for those months to 19.3 percent, and during July and August, normally the heaviest months during *keremt*, rainfall was 10.9

TABLE 3.1. Deviations as a percent of annual mean rainfall in Bati, Rike, and Qobo

	1972	*1973*	*1981*	*1982*	*1984*
Bati	99.4	77.4	87.3	103.4	37.4
Rike	80.1	ND	141.0	155.4	52.3
Qobo	ND	ND	119.3	98.2	19.1

Source: Calculated from Dejene, *Environment, Famine, and Politics in Ethiopia*, 56, 57, and 84, tables 3.3, 3.4, and 4.2.
ND = no data.

percent and 16.6 percent of the long-term mean. Only out-of-season spikes in May and December, rain that was either useless from an agricultural standpoint or even damaging, allowed Bati's annual total to approach even 37 percent of the long-term mean. Similarly, the total rainfall in Rike for 1984 was a little over half (52.3 percent) of the long-term mean. The *balg* rains in Rike were 18.8 percent, 67.3 percent, and 13.2 percent in February, March, and April, respectively, of the long-term mean for those months. As was happening in Bati, in Rike an unseasonably high amount of rain fell in May (234.1 percent of the mean). During *keremt*, July rainfall was 44.6 percent of the mean; and in August rainfall declined to 9.4 percent of the mean. Only the major spike in May allowed Rike's annual total to reach 52.3 percent of the long-term mean.

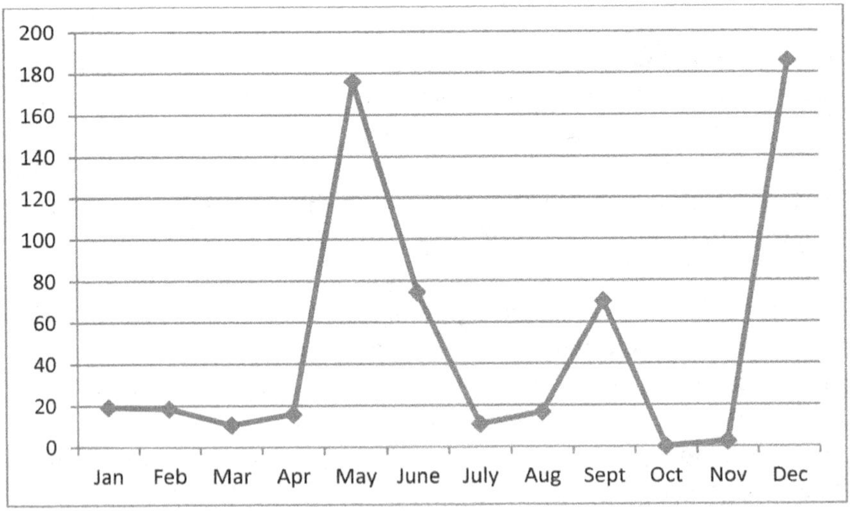

Figure 3.3. Bati 1984 rainfall as percent of monthly mean, 1968–1988

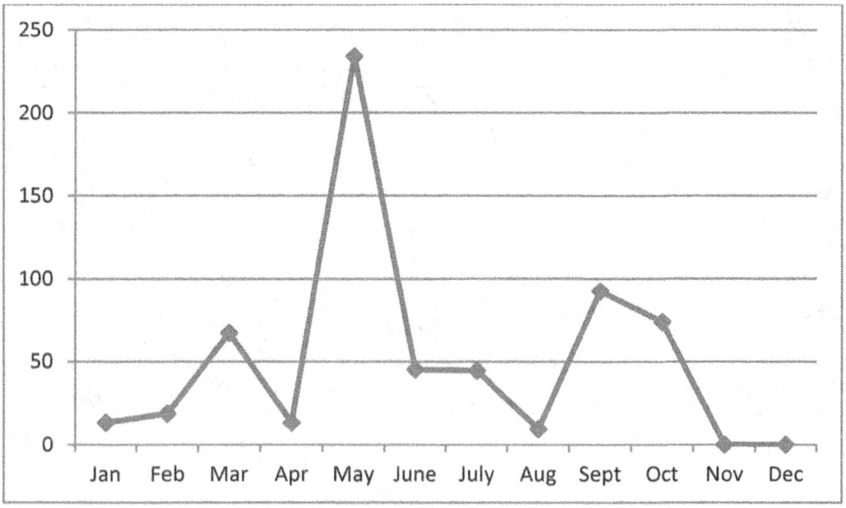

Figure 3.4. Rike 1984 rainfall as percent of monthly mean, 1968–1988

If Bati and Rike in 1984 demonstrate the coincidence of agricultural and meteorological drought, Bati in 1982 provides an example of agricultural drought during a year in which there was no meteorological drought. In that year Bati's annual rainfall was 103.4 percent of the annual mean rainfall for 1968 to 1988, but this rain fell almost entirely out of season. Table 3.2 provides the numbers that figure 3.5 illustrates. January, October, November, and December experienced rainfall far above the long-term mean, while during the agriculturally important months of February and March no rain fell at all. Nor did any rain fall in June. In the key *keremt* months of July and August, rainfall was 38.9 percent and 74.9 percent of the long-term mean for those months.

Although there is no controversy about the occurrence of both drought and famine in 1984, controversy does surround the role of drought in the origins and unfolding of that famine. Penrose claims that "the development of a disastrous famine in 1984 must . . . be seen against the succession of poor rains and inadequate harvests in the four years after 1980." Her account is similar to Workineh's, which we considered already, and was held by most relief agencies operating in Ethiopia at that time. According to Penrose, following recovery from the drought of the early 1970s, drought-induced stress in northern Ethiopia began again as early as 1979.[41] This is consonant with the public pronouncements of the Relief

TABLE 3.2. Bati 1982 rainfall as percent of monthly mean, 1968–1988

Month	*Rainfall (% of monthly mean, 1968–88)*
January	267.4
February	0
March	0
April	19.5
May	39.2
June	0
July	38.9
August	74.9
September	100.3
October	436.3
November	658.9
December	348.6
Annual	103.4

Source: Dejene, *Environment, Famine, and Politics in Ethiopia*, 57, table 3.4.

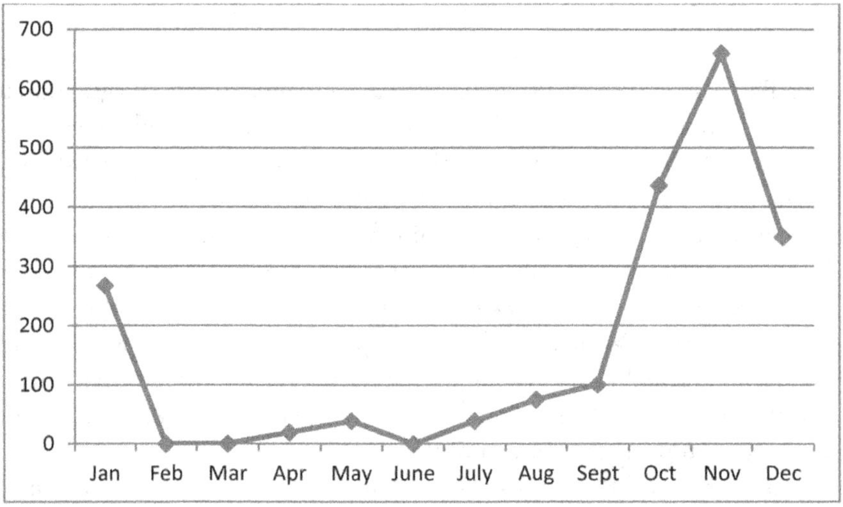

Figure 3.5. Bati 1982 rainfall as percent of monthly mean, 1968–1988

and Rehabilitation Commission (RRC). However, on the basis of a close reading of RRC reports, the behavior of changes in the price of grain and his reading of the climate data, de Waal challenges Penrose and charges that those "who wish to explain the famine in meteorological terms are impelled to fall back on either convoluted arguments or blatant untruths."[42] This is a harsh judgment.

De Waal dismisses 1982 as a year contributing to the build-up to famine and characterizes it as having "the largest ever harvest" and as a "bumper year." Most of the rainfall data do present 1982 as a year above the long term. The graphs and national data about "Cereal Production per annum 1974–1986" and "Agricultural Food Production per capita" as presented in the discussion below also show a spike in 1982 against a general downward trend from 1979. This is apparent from figure 3.2, where 1982 is a notable exception to the general deficit in annual rainfall, which marked 1978 to 1984. The charts of annual rainfall for Bati, Kobo, and Rike (figs. 3.6–3.8) all show 1982 as a year above the mean.

As we have already seen, the case of Bati in 1982 is deceptive, because although it experienced no meteorological drought that year, it did suffer from agricultural drought. On the basis of information covering the years 1965 to 1999 from more than twenty sites in eastern and northeast Ethiopia (which cover the entire area of interest here), along with a number of stations in the desert lowlands east of Wallo proper, Zewde and Lamb established that in 1982 and 1983 *keremt* was unusually late in setting in, and both years had growing season precipitation well below the thirty-five-year mean.[43] Regional variation is also evident in the data presented by Woldeamlak and Conway in their study of rainfall variability in the Amhara region. In 1982 three localities—Kombolcha, Dabra Marqos, and Kamisse—received annual rainfall above the long-term mean from 1981 to 2003, and three other localities—Bahir Dar, Dessie, and Gondar—were marked by deficits. Of these six sites, three are in Wallo. Two of the three received above average rainfall; one received less. Combining this information with the information from Bati and noting that 1982 and 1983 might have been years of agricultural drought, even in the absence of meteorological drought, it would seem cavalier to dismiss the possibility that 1982 could have contributed to accumulating drought. This would bring us in line with Dessalegn Rahmato's informants, who told him that serious food shortage began in lowland Wallo (Warrababo and Tahuladare) in 1982, when the *balg* and *keremt* rains proved inadequate. They also testified to shortages of rainfall in *keremt* 1983, which fed into the rainfall failures and widespread famine of 1984.[44]

Yet another vexing question, raised by de Waal and by Glantz and Workineh, is just how much drought does it take to make a famine? We do not seem to be close to finding an adequate answer. Woldeamlak and

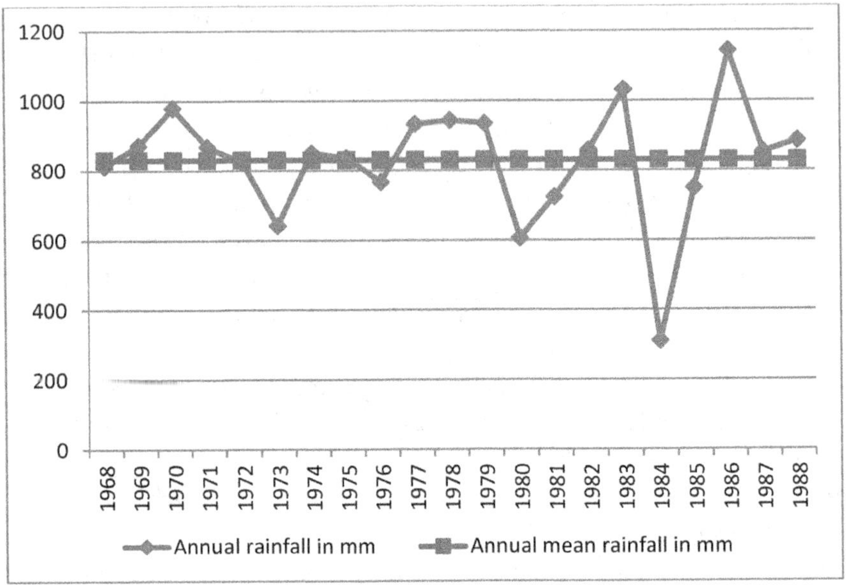

Figure 3.6. Bati annual rainfall, 1968–1988

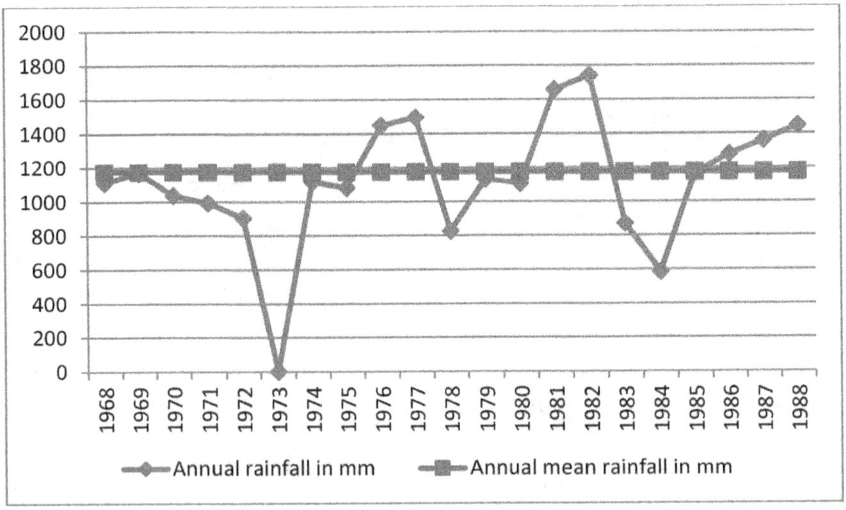

Figure 3.7. Rike annual rainfall, 1968–1988

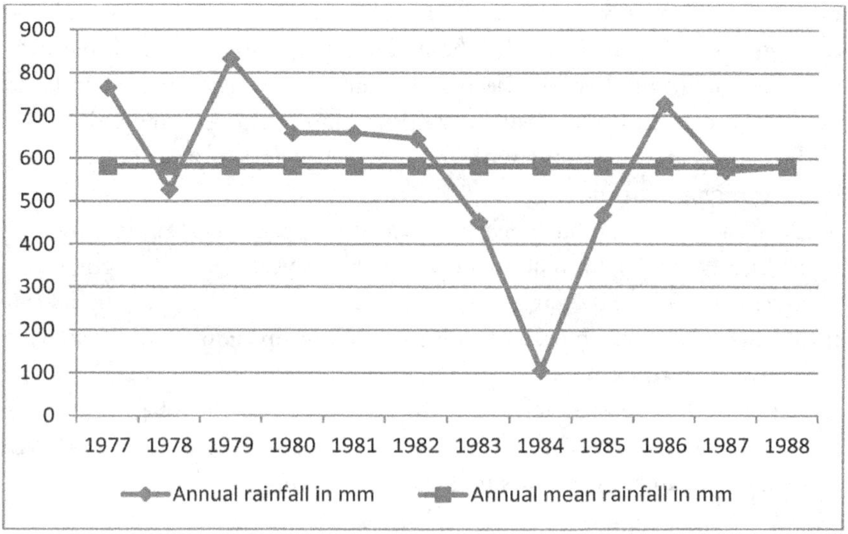

Figure 3.8. Qobo annual rainfall, 1968–1988

Conway cite a report by the International Food Policy Research Institute to the effect that "a 10 percent decrease in seasonal rainfall from the long-term average generally translates into a 4.4 percent decrease in the country's food production."[45] It is unlikely that the relationship between rainfall and food production would maintain a simple arithmetical progression. Were this so, a 100 percent failure of rains would produce only a 44 percent decline in food production. Moreover, this progression does not seem to provide for the cumulative effect of a succession of below-normal rainfall years.

Glantz and Workineh draw attention to the fact that while meteorological conditions might not change, land users can increase drought vulnerability by moving into lands that are climatically more marginal than those from which they came. Settlement has not been static in the area under study.[46] All through the eastern highland plateau, farmers having been moving both higher and lower into lands not previously exploited by their parents. Such movement is much remarked on in the literature, although it has been the subject of few dedicated studies. In the Simien Mountains and elsewhere, farmers have been moving upslope since the 1950s.[47] Here they may be taking advantage of rising temperature, which raises the altitude at which crops become vulnerable to frost. They have also been moving downslope to the east into *qolla* lands not farmed by earlier generations.[48]

Here the presumption is that they are being pushed out of ancestral lands by rising population pressure. These *dagga* lands are indeed climatically more marginal, and they have been conspicuously prone to drought-induced famine. In 1973 and 1984, the eastern downslope lands in northern Shawa, Wallo, and Tigray were hit with famine earlier and more intensely than lands at higher altitudes.

De Waal appears to have been correct to the extent that nationally 1982 was a good year for rainfall and harvest. However, in the drought-prone northeast it was a year of agricultural climate stress, which, playing its role in a five-year decline in agricultural production preceding 1984, may very well have intensified the effect of the grievous drought of 1984. De Waal's rejection of climate as a factor contributing to famine in 1984, arises from his conviction that the real culprit was the war the Ethiopian government was waging against insurgents in Tigray province.

War

Northeast Ethiopia has been vulnerable to drought-related famine since 1965. The years 2002–3 witnessed the most grievous famine since 1984, with harvest loss running up to 80 percent in some places.[49] The country was at peace in 1965, 1973, and in 2002–3, so warfare cannot explain the country's ongoing vulnerability to famine. Yet the famine of 1984 did coincide with civil warfare, which at the least must have intensified the effects of the famine.

Following the revolution of 1974, Ethiopia found itself involved in three major wars—against Somalia and in Eritrea and Tigray.[50] The Somali war of 1977–78, fought on the plains of the Ogaden and the highlands around Harar, far to the south and east of Wallo, does not bear so directly on our story as do the others, which were fought in some of the famine-affected areas of northeast Ethiopia. The origins of the Eritrean war lie in an armed insurrection against the government of Haile Selassie I in the late 1950s, directed to achieving independence. With Addis Ababa distracted by revolutionary politics and war with Somalia, the Eritreans, led by the Eritrean People's Liberation Front (EPLF), established a high water mark in 1977–78, driving the Ethiopian army from all but a handful of garrisons in Eritrea. The tables turned when the Ethiopian armies, which had driven the Somalia army from Ethiopia with Russian and Cuban support, were redeployed against the Eritreans in 1978. In 1982 the Darg, deploying heavy artillery, tanks, and jet aircraft, launched a series of large-scale campaigns that combined political and military offensives but failed to crush Eritrean resistance, which retreated to what proved to be an impregnable stronghold at Naqfa in northern Eritrea and into the countryside.[51]

Insurrection in Tigray was a product of the revolution itself. In the mid-1970s a number of groups vied to establish an armed presence there in opposition to the Darg. The TPLF emerged supreme and by 1980 was firmly established in the province, with a clandestine political presence in the province's towns and a guerrilla presence in much of the countryside. Its objective was the autonomy of Tigray province and its Tigrinya-speaking population. As its ambitions grew, they came to include the reconstruction of the entire country on the basis of ethnically defined regional states.[52] The Darg was slow to realize the threat posed by the TPLF. However, in 1980 it launched a series of military campaigns, which ran through 1985.

At the time of the outbreak of famine in 1984, the Darg appeared to hold the military upper hand in both Eritrea and Tigray. The founding of the Workers' Party of Ethiopia (WPE) in September 1984 and the People's Democratic Republic of Ethiopia in September 1988 finally institutionalized the revolution.[53] Moreover, the Darg controlled the great bulk of famine relief, which the RRC in association with a plethora of NGOs distributed through a network of relief camps. Appearances were deceiving. The TPLF and EPLF achieved a close working relationship with each other in the face of the intense Ethiopian government campaigns of the early 1980s, but the cumulative pressure of these campaigns, along with the impact of the famine and long-standing tensions between the two movements led to a rupture in 1985, which was not mended until 1988. In March 1988 a Darg army suffered a stunning defeat at Afabet in Eritrea at the hands of the EPLF, and the Darg's military dominance inexorably unraveled over the next three years. A TPLF victory at Enda Sellase in western Tigray in February 1989 led to the collapse of the Darg's presence in Tigray, with the TPLF occupying the provincial capital, Maqale, that same month. In May the commanders of most of the Darg army attempted a coup d'état, which, although it failed, proved another step in the disintegration of the regime. In February 1990 Massawa fell to the EPLF and the TPLF army invaded Wallo and Gondar. In March 1990 the government in Addis Ababa announced the abandonment of the socialist command economy and the political monopoly held by the WPE. Too little, too late. In May 1991 the EPLF occupied Asmara, the TPLF occupied Addis Ababa, and Mengestu Haile Mariam fled to exile in Zimbabwe.[54]

Military operations undoubtedly exacerbated the effects of drought from which Eritrea and Tigray, like Wallo and Gondar, suffered.[55] Alex de Waal argues that they were responsible for the famine: "The principal cause of the famine was the counter-insurgency campaign of the Ethiopian army and air force in Tigray and north Wollo during 1980–85." He writes of five different factors, three of which bear on the origins of famine: military offensives against TPLF strongholds from 1980 onward, bombing of markets in rebel-held areas, and restrictions on movement and trade.[56] First, he references the sequence of military campaigns between 1980 and 1985 launched by the Ethiopian government against the TPLF, which "destroyed grain stores, burned crops and pastures, killed livestock, burned houses, enforced the collection of taxes and contributions including 'arrears,' and displaced about 80,000 farmers." Second, he writes of aerial bombing of truck traffic and of rural markets in TPLF-held territories. Bombardment killed hundreds of people and, more important, disrupted both agricultural production and the normal flow of goods. This leads directly to his third point, the restriction of commerce and migration. Checkpoints and a network of garrisons stifled the flow of grain from surplus-producing western Tigray to famine-affected eastern parts of the province: "Counter-insurgency restrictions on the grain trade were instrumental in starting the famine, and played a key role in maintaining it."[57] His fourth and fifth factors are forced relocation of farmers and the manipulation of relief programs. He concedes that neither played a causal role in the famine, since both followed the famine's outbreak, whatever their effect on famine conditions and the conduct of the war.

De Waal's case requires that we view the famine of 1984–85 as largely confined to southern Tigray and adjacent parts of north Wallo and the impact of drought as exaggerated in the area it affected and in its intensity. Koram, the most notorious of the relief camps, was indeed in northern Wallo, on the eastern escarpment close to the border with Tigray, but the famine of 1984 affected large parts of Wallo, eastern Gondar, and northern Shawa as well, where the impact of war was modest at best.[58] De Waal does acknowledge the widespread failure throughout Wallo, eastern Tigray, northern Shawa, and eastern Gondar of both the *balg* and *meher* rains in 1984, but challenges the common understanding that the effect of these droughts was intensified by drought in 1983 as well.

While de Waal's view has not gained general acceptance, it does draw attention to the interaction of war and famine, particularly in Tigray and

Eritrea. The scale of operations alone would suggest that they bore heavily on the countryside. Between 1974 and 1991, over a million soldiers served in the Ethiopian army, some in campaigns in the Ogaden and elsewhere in southern Ethiopia, but a large proportion of them in Eritrea and Tigray, where the insurgents had large numbers under arms. In waging war, the Ethiopian government incurred a foreign debt of over US$4 billion and had a defense budget that was twice the combined budgets for education, health, agriculture, and industry.[59] In 1987, on the eve of their breakthrough campaigns, the TPLF and the EPLF had somewhere between 75,000 and 105,000 guerrillas in the field. Gebru Tareke, drawing on Ethiopian government archives, estimates that the wars, exclusive of famine, took the lives of 800,000–1,000,000 Ethiopians and Eritreans. Many Ethiopians went into exile, fleeing famine and warfare in addition to the revolution. In 1988 the Ethiopian government estimated that there were some 870,000 Ethiopian refugees in Sudan and Saudi Arabia alone. The loss of the labor of 1.6–1.8 million people must have negatively affected the capacity of rural households to respond to developing drought.[60]

De Waal's account of the many ways military operations bore directly on farmers, often intensifying the effects of drought, is supported by other sources. A survey of farmers carried out in 1987 by the EPLF and its relief arm, the Eritrean Relief Association, indicated that thousands of young men were conscripted, agricultural lands destroyed, food confiscated, animals stolen, people killed and imprisoned, and homes destroyed.[61] In the Sudan, Clay and Holcomb interviewed Ethiopian refugees from resettlement camps, most originally from Tigray or Wallo. Their informants relate the same appropriation of livestock and tools, burning of grain stores and houses, bombing of villages, and killing of civilians.[62] Although it is impossible to verify these claims, they do echo the hardship that warfare had wrought a hundred years before at the time of the Kefu Qan.

Government Policy

As was the case with warfare, "socialist" economic policy was not in play in 1973–74 and had been dropped in 1990, before the fall of the Darg, so it cannot explain Ethiopia's long-standing famine vulnerability. This does not mean that government policies had no effect on the course of famine in the 1980s. By the late 1970s the Darg was fully committed to fashioning Ethiopia's economy after the command economies of the Soviet Union and Eastern Europe. Its long-term strategy for increasing agricultural productivity was self-consciously socialist—the creation of state farms and producers' cooperatives to which it channeled its investment in agriculture. Its attitude toward the peasant sector was patronizing and extractive. A review of agricultural development in Ethiopia under the Darg noted that "some 85%, 50%, and 79% of total agricultural credit, fertilizer and improved seeds respectively go directly to socialist enterprises (state farms and producers' cooperative) which, up to 1985, have jointly accounted for only 5% of the total cultivated land and 4% of the national crop production."[63] Over the next six years, to 1991, this misallocation was not changed. However, although the misconceived strategy marked little change from the previous state of affairs under Haile Selassie, which had also seen little government investment in agriculture, it was accompanied by a steady decline in cereal production and food production on a per capita basis.

In 1975–76, following the revolution and the nationalization of agricultural land, cereal production had risen rapidly, then maintained its level to 1979, when, as figure 3.9 shows, it fell steadily, with one upward spike in 1982–83, to collapse in 1984–85 and only modestly recover in 1985–86. The evidence for the production of all agricultural sources of food closely parallels overall cereal production, as figure 3.10 shows. It took four years after the famine for agricultural food production to return to the level of the famine's eve, and in retrospect the early rise in 1975–76 was interpreted as recovery from the famine of 1973–74 and the result of favorable rains.[64] The relationship between the decline in the availability of food and famine vulnerability seems obvious, especially in light of Ethiopia's failure to expand its exports to find other resources that would have allowed it to import food on a sustained basis.

In addition to misdirected investment, Darg policies, which extracted resources from farmers tilling 95 percent of Ethiopia's agricultural land, contributed in some degree to discourage production or weaken the farmers' capacity to cope with environmental challenges. Under the Darg farmers continued to contribute in direct taxation much the same proportion (about 5 percent) to government revenue as they had done under Haile Selassie, but the volume of that revenue swelled by 156 percent in the decade following 1973–74. The Darg levied a land use fee and an agricultural income tax, which apparently amounted to about B20 per annum per capita. Farmers had to pay much more under the Darg than they had under Haile Selassie to local organizations: annual membership fees to their peasant associations and in support of the women's and youth organizations

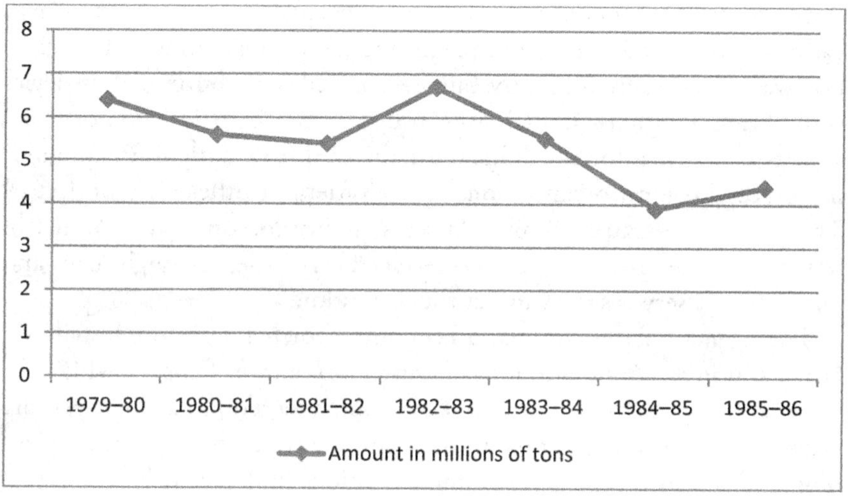

Figure 3.9. Ethiopia cereal production by year, 1979–1986
Source: Clapham, *Transformation and Continuity in Revolutionary Ethiopia*, 166.

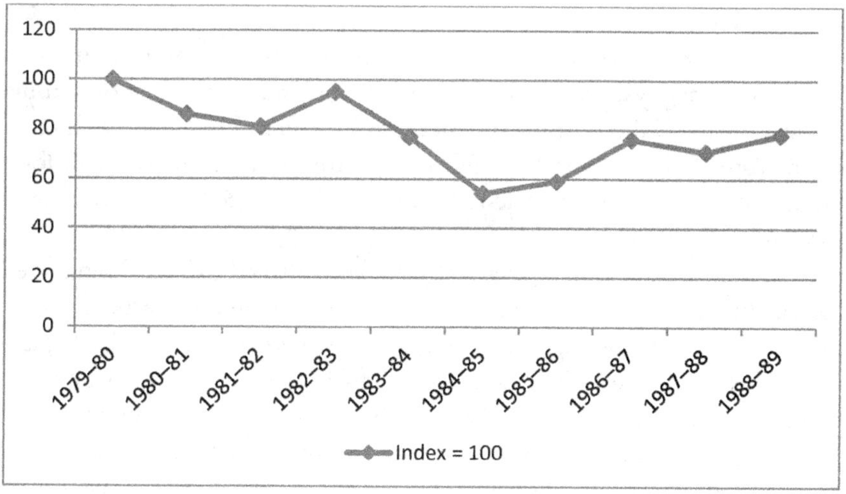

Figure 3.10. Agricultural food production per capita, 1979–1980
Source: Hansson, *The Ethiopian Economy 1974–94*, 166.

and a succession of "public fund-raising campaigns" in support of the Darg's various wars, a rural literacy campaign, and famine relief in 1985. Participation was theoretically voluntary, but few escaped contributing. These levies could bring a farmer's annual taxes to B90 a year. Unpaid labor, of many as forty days a year for all adults, was redirected from agriculture to building offices for local organizations and a variety of other civic projects.[65] The loss of these assets limited farmers' ability to cope with famine; in return for these lost assets, farmers received very little, although roads and public health services grew under the revolution.

The agency of the command economy, which bore most heavily on farmers, was the Agricultural Marketing Corporation (AMC), which the Darg created in 1976 to control trade and channel food to its high-priority areas—the army, other national institutions, and the towns. Central planning authorities established commodity prices, and the AMC set quotas that the different peasant associations had to meet by selling specific grains to the corporation at corporation-determined prices. A parallel market operated by private traders was increasingly squeezed in favor of the AMC. The free movement of cereals from province to province was steadily limited. Checkpoints were set up outside all the principal towns to curtail private grain movement.[66] Distrustful of the market in principle, and convinced that farmers diverted too much energy and attention from field work to attending markets, the Darg insisted that all markets be held only on Saturdays.[67]

Underlying all these measures was the obtuseness and insensitivity with which the Darg system operated. It levied the famine relief tax on famine victims, just as it did on those unaffected by famine. The AMC took more grain from Wallo from 1980 onward than the province received in food aid in 1983–84—59,500 metric tons compared with 36,600 metric tons of food aid.[68]

A number of additional programs have been alleged to have discouraged agricultural productivity and increased famine vulnerability: the 1975 land reform, formation of producers' cooperatives, resettlement, and villagization. Most plausible is the land reform, which had ended landlessness and was popular in many respects. During the late Haile Selassie years, a variety of experts had advocated land reform, hoping it would unleash latent productivity. It did not do so. The actual reforms, together with the formation of peasant associations, had the effect of restricting movement, since farmers now had rights to land only in their original peasant association.

No new land was added to that already under the control of the associations, which, with curtailment on movement from the peasant association into towns and a rate of population increase of 2.6 percent a year, meant a slow reduction in the land available for each farm family.[69] This reduction was increased by various government actions that removed land from production—for conservation, local offices, and a host of other functions.

The contribution of cooperatives, resettlement, and villagization to discouraging agricultural production and increasing famine vulnerability is more tenuous. The 1975 enabling proclamation for the peasant associations announced that they were "to induce and organize peasants into co-operative farms." There were two different kinds of cooperatives—service cooperatives and producers' cooperatives. The former, much more common, purchased and distributed goods. The producers' cooperatives were to engage in common productive activities and were the goal of the 1975 proclamation. Clay and Holcomb claim that famine resulted "primarily from government policies that have been implemented in order to accomplish massive collectivization of agricultural production."[70] Given that in 1984 only about 6 percent of the peasant associations in Wallo had been reorganized into cooperatives and that the program was little implemented in famine-afflicted Tigray and Eritrea, its contribution to creating famine must have been marginal.[71]

Resettlement and villagization equally played no role in generating famine in northern Ethiopia in the mid-1980s. Resettlement was a response to the famine, and villagization is a program that was launched nationally in October 1985, after the worst of the famine had passed. Resettlement entailed the transfer of several hundred thousand farmers to ostensibly more fertile, less-stressed parts of the country, principally in western Ethiopia, toward the border of the Sudan. A plausible case has been made that resettlement may have introduced famine into the reception areas, but this has nothing to do with the genesis of famine in northern Ethiopia.[72] In Wallo the relationship between famine and the formation of cooperatives may have worked in the opposite direction—famine inducing a greater willingness to enter cooperatives. In the country as a whole, the formation of cooperatives between June 1984 and July 1985 increased by 75 percent. In Wallo it rose 515 percent over the same period. Clapham attributes this most probably to the fact that members of cooperatives were exempt from the resettlement program, although leverage the government gained through the control of famine relief may also have contributed.[73]

Villagization arose in Bale in southeast Ethiopia as a security measure and took on a life of its own. It was picked up by the national state following Mengistu Haile Mariam's visit to Hararghe in June 1985. It became the object of yet another government campaigns in October that same year. The earliest reports of villagization in Gondar and Wallo were in July 1987, and I witnessed the process in full swing in Gondar in February 1988. Ostensibly the national program aimed to concentrate rural populations for the better provision of amenities such as water pumps, clinics, and schools. More likely is that its popularity in governing circles came from the greater control over rural populations, which it facilitated, and the groundwork, which it laid for forming future producers' cooperatives. Its contribution to original famine vulnerability was negligible, taking place well after the famine crisis had passed. How much if at all these factors actually constrained agricultural production is impossible to say, but they did constrain the ability of farmers to respond creatively to food shortages as they emerged and intensified.[74]

Coda

Apocalyptic visions of a highlands stripped of its vegetation and soil fueled major interventions into landscape management. These visions and interventions were not the product of the lived experience of Ethiopian farming men and women, but through enticement and pressure they provided the labor to make them a reality. They planted millions of trees. They created thousands of kilometers of bunds, check dams, and terraces for tree planting. What they thought of the interventions they were able to express openly in 1991 and 1992 during the transition from the People's Democratic Republic of Ethiopia to the Federal Democratic Republic of Ethiopia, when they destroyed many of the interventions.

Scholarship has harshly judged the conservation programs of the 1980s. Yeraswork characterized them as "20 years to nowhere" and Dessalegn wrote of "littering the landscape."[75] Theirs will not be the final word. The authorities who succeeded the Darg pursued similar landscape goals of soil conservation, afforestation, and enclosure and pursued them often in a similarly authoritarian manner. The Ethiopian landscape of the 1990s and twenty-first century gives some evidence of an increasing number of trees and the adoption of conservation measures.[76] The extent to which these measures have been incorporated into their regular practices by farmers as opposed to being imposed by external authorities remains to be established.

The vision of a physically bankrupt highland environment succeeded the famine of 1973, and was well established at the time of the famine in 1984. The vision reinforced the sense of crisis, which quite reasonably accompanied the famines. But in what sense was the crisis environmental? It was environmental in that Ethiopian farmers found themselves having to cope with drought unprecedented in their experience. It was not necessarily a landscape crisis, but it was an agrarian crisis. The extent to which it was a crisis of political economy is less clear. The great famines of 1888–92, 1973–74, and 1984–85 occurred under very different regimes—Menilek's premodern state, Haile Selassie's would-be capitalist state, and the Darg's People's Republic with its command economy. The common agrarian thread running through these divergent arrangements of political economy is a farming system where control of production remained in the hands of the cultivators and in which the state's relationship with the farmer was predatory.

The landscape visions animating the conservation programs of the 1970s and 1980s entailed projections and inferences rooted in the Euro-American past. They were not fabrications, but they were poorly grounded in Ethiopian landscape history, a subject chapter 5 explores. Chapter 6 delves into the experiences of Ethiopian farming men and women who lived through the landscape changes discussed in chapter 5. First chapter 4 looks at the scholarship around farmers and responses to famine.

4

Farmers Cope

THE RESEARCH CARRIED OUT in the mid-1980s by the Soil Conservation Research Project (SCRP), directed by Hans Hurni, and by the Ethiopian Highlands Reclamation Study (EHRS) was deeply animated by the degradation narrative, which entailed recent, dramatic landscape change brought about by the dysfunctional premodern agricultural practices of Ethiopian farmers. The most conspicuous shortcoming of these studies, and the interventions to which they gave rise, was their failure to adequately incorporate the views of Ethiopian farmers, despite explicit recommendations in some studies that this be done.[1] It is unclear if any of these recommendations had been considered by either the SCRP or EHRS. This did not stop researchers from looking for a deeper understanding of the needs, interests, and coping strategies of Ethiopian farmers. Conspicuous among these researchers was Amartya Sen.

Farmers, Markets, and Famine

Amartya Sen's book *Poverty and Famines* revolved around the role of markets and the evidence they provided for the social impacts of famine and for how local society responded to its challenge.[2] His work drew on the work of others, most notably Seaman and Holt.[3] It launched a debate about whether the Ethiopian famine of 1973–74 was primarily caused by food availability decline (FAD) or by a deficit in the availability of what Sen calls "entitlements" (food entitlement decline, or FED). One summary of this distinction notes that "FAD concentrates on the total availability of food in a geographical region, while FED focuses instead on the distribution of available food between different groups of people."[4] That there was a collapse in the availability of food in Wallo seems obvious given the failure of the main harvest in 1972 and the *balg* rains in 1973. However, Sen could find no evidence of this FAD in national data concerning food production for 1972 and 1973, so he asked the question, why was food not imported from other parts of Ethiopia, where drought and crop loss were indiscernible? A related question that might help answer the first question was, who were the famine victims?

Sen was convinced that the key to understanding the unfolding of famine in Wallo was the collapse of markets for food in the province during the famine months. There appeared to be no evidence of significant movement in the prices of foodstuffs, instead of the large rise in demand, which one might expect. Why did the sudden loss of harvest not express itself in attempts to buy food? According to Sen, harvest collapse deprived Wallo farmers not only of their immediate source of food but also of the wherewithal to generate cash to buy food. There was no significant rise in the price of food because there was no increase in the real demand for it as felt by the market. Wallo farmers lacked "market command." Sen dismissed arguments that the inadequacy of roads in the province hampered importing food and contributed to market shortages. He pointed out that the province was transected by two national highways and the admitted paucity of back-country roads spoke only to intraprovince distribution, not to imports from the national market. Moreover, he argued, there appeared to have been movement of food out of Wallo during the famine months, a further suggestion that there was little market demand for food within the province and that there were no blockages to transportation.[5]

The famine struck along the main north–south highway running from Addis Ababa to Asmara, affecting more the areas to the east than to the west of the highway. The road itself runs through *dagga* and *wayna dagga* zones with the land to its east increasingly falling away to *qolla*. Three districts in Wallo were hit particularly hard: Raya and Kobo, Yaju, and Ambasäl (of which Tahuladare is a part). They provided over three-quarters of the population in the relief camps. The highland districts in western Wallo were only modestly affected. Pastoral and farming peoples were affected by the famine. A much higher percentage of pastoralists suffered than did farmers, who, however, because they greatly outnumbered the former, constituted a large majority of famine victims. Although the farming population was not highly differentiated internally, famine emphasized what differences did exist. Not all rural folk suffered equally.

An analysis of famine refugees in the relief camps revealed that they disproportionately consisted of women and children from nearby farming communities, relatives and dependents of farm families, farmers whose holdings in land were particularly small, and tenants.[6] Townsfolk were also represented by day laborers, "women in service occupations," and the like. Artisans were disproportionately numerous in the shelters. To Sen, these groups of people reflected the coping strategies that many farmers, faced with crop failure, adopted. Many families broke up for what was intended to be a temporary arrangement—the men going in search of wage labor and the women and children going to the camps, where they could get food. Other households stayed intact and in situ only by shedding all but their core members. Many of their attached relatives had shared the household labor and when the value of what they contributed became less than what their retention was going to cost, they were let go. The households with the least amount of land had been dependent on more prosperous neighbors for rental and other arrangements, which provided some downward transfer of resources. Many tenants suffered from the inflexibility of their landlords, who refused to adjust rental arrangements, although some did just that. Most towns were embedded in their surrounding countrysides, and losses in the rural economy were directly felt in the towns.

Seaman and Holt revisited the Wallo famine in 1980, comparing it with contemporary famines in Bangladesh and the West African Sahel. They disagreed explicitly with Sen, arguing that there had indeed been a rise in the price of food, "quite out of proportion to the supply of food in the area," a rise that magnified "the extent of starvation."[7] Sen had put too much emphasis on price information from the Dessie market, they argued, and too little on parallel information from smaller markets, closer to the heart of the famine. They had found their evidence in a fivefold increase in food prices in Alamata, in northern Wallo, where the famine struck with particular severity. This rapid rise saw a simultaneous "fall in the cash value of the assets of those dependent upon the market for food." While food prices fairly quickly returned to normal in Alamata, Seaman and Holt suggest that the increase may have rippled out from this epicenter. They attribute the increase to three main factors:

(1) the failure of supply from drought;
(2) an increase in demand, as farmers who had failed to meet their subsistence needs sold and mortgaged assets to buy food;

(3) likely the withholding of surplus by larger farmers, and possibly continued exports of food from the region.[8]

Farmers faced with crop failure had a number of options to get the cash with which to buy food. They could borrow the money, or they could sell their land or labor. With so many of their neighbors in the same position, most farmers found competition for loans intense and the value of their land or labor driven downward by competition. Nonetheless, it appears that by borrowing and liquidating assets, many were able to stave off the worst effects of the famine, at least in the short run, and hence delay their complete lack of market command.

In its insistence that this process cannot be understood simply with reference to the collapse of harvests in 1972 and 1973, in its use of market data, and in its demand for a close look at the behavior of farmers during the onset of famine, the debate stimulated by Sen's book did advance our understanding of the unfolding of famine in Wallo in 1972 and 1973. One of its striking features is how its silence highlights the absence of an informed economic history of Wallo, which in turn reflects the slender corpus of work on the economic history of Ethiopia in general.[9] I recall just one reference in the debate to Wallo's place in the Ethiopian national economy prior to 1972. This is the claim by Seaman and Holt: "In normal years, Wollo is a major exporter of surplus crops to the capital, Addis Ababa, and to the northern provinces."[10] This may or not have been true when the remark was made in 1980. My own foray into understanding Wallo's participation in the national economy suggested that in the years following the famine, it was slight indeed.

The early phase of the debate about the Wallo famine of 1972–73 hinged on one data set—the market prices of food in northern Wallo, gathered firsthand by Seaman and Holt in 1973 at Korem, Alamata, and Kobbo: "Prices for staple grains rose rapidly during the first part of 1973, so that by the beginning of the rains, in the middle of the year, they had nearly doubled. They fell again with startling rapidity at the end of the rains, so that in September they had nearly returned to pre-famine levels."[11] In a later statement, they noted that price information for Wallo is "only fragmentary" and went on to note that prices must have risen rapidly in late 1972: "Early in 1973, prices in Alamata . . . had risen to 100 E$/100 kg, approximately five times normal; prices of sorghum and tef, the usual staples, fell dramatically, but briefly, at times when the government distributed grain (Woyzero

Kebedech, 1980). At the same period, prices in the provincial capital, Dessie, were reportedly lower (54 E$/100 kg), but still much higher than the seasonal norm."[12] Information about the market prices of food grains did expand with "slightly better data from the famine of 1982–85."[13]

The famine of 1972–74 helped put in place institutions that were charged with generating information for anticipating and preventing future famines, the national Relief and Rehabilitation Commission (RRC) foremost among them. By 1984 the Central Statistical Office was providing on a monthly (albeit patchy) basis average retail prices of major food grains at the level of the *warada*. The Save the Children Fund provided financial support to the RRC to provide weekly average retail price data from several sites in Wallo, one of which was Koram. The Ministry of Agriculture issued data on sales of livestock, a crucial element in the economy of Wallo, not to mention the country at large.[14]

RRC data register drastic declines between 1983 and 1984 in the production of food in most of the *awraja* of Wallo. Table 4.1 lays out these data. Although our source describes these data as spotty, the case for a rise in the price of food grains accompanying the 1984–85 famine seems strong. Evidence of sharp rises in prices starts as early as 1982. In June, July, and August of that year, prices for all cereals apparently rose in Kombolcha and Lalibala. In July 1983 retail prices for wheat and *tef* in drought-affected Tigray had climbed to 200 birr/quintal against a more normal figure of 90 b/q. In Koram food prices rose sharply from June to October, peaking in late September at "very high levels": *tef*, 160 b/q; wheat, 152 b/q; sorghum, 112 b/q; and barley, 105 b/q. In October prices dropped to "normal" levels, only to rise again. The prices for individual food grains maintained their differentials, rising and falling in tandem. Cutler summarized these data: "as a generalization it seems that where crop failures are reported to be at their worst, prices are highest."[15]

Later discussion has generally agreed that Seaman and Holt's discovery of a price rise in 1972–73—which Sen marginalized but the plausibility of which was boosted by the price rises that occurred in 1983 and 1984—was a serious challenge to Sen's view that Wallo farmers suffered an immediate collapse in market command.[16] Later commentators have argued that Sen downplayed the extent to which Wallo suffered crop loss in 1972 and 1973 and minimized the effectiveness of farmer adaptation to the collapse of production. Measures such as the sale of assets and adjustment in consumption—shifting demand from more prestigious cereals like *tef* to

TABLE 4.1. Food production declines and numbers affected in Wallo, 1983–1984

Awraja*	*Average percentage change in production from 1983 to 1984*	*Number of people affected by food shortages*		*Differences in number of people affected*
		1983	*1984*	
Rayana Kobo	–90	63,950	169,790	+105,840
Yäjju	–85	44,680	133,990	+83,310
Ambasäl	–80	85,600	195,550	+109,950
Dessie Zuria	–65	11,800	65,660	+53,800

Source: Adapted from Baulch, "Entitlements and the Wollo Famine of 1982–1985," 198, table 3. Baulch provides data for all eleven *awraja* of Wallo. I selected the four *awraja* of greatest pertinence to this study. Baulch's source: *RRC Crop Dependent Food Supply System Report*, 1984.

less prestigious ones, like sorghum—seem to have had considerable impact in lessening the immediate impact of drought and famine. At the same time, Sen's insistence on an entitlement analysis, one that is sensitive to social differentiation and gives considerations of demand equal weight with considerations of supply, has received wide endorsement.

From Poverty to Famine

James McCann's *From Poverty to Famine* provides some of the historical background so conspicuously missing in the SCRP and EHRS studies and the Sen debate. It is a rural social history of the northern Wallo region of Lasta, which was a source of out-migration for much of the twentieth century. McCann, funded by the US Social Science Research Council and the Fulbright-Hays Program, carried out his research for the book in Ethiopia in 1982 and for Oxfam America in 1985. *Poverty to Famine* rests on fieldwork, documentary and archival sources, and interviews with informants from Lasta. Concrete datable evidence from Lasta is scarce, hence much of the argument proceeds by inference and analogy drawn from developments in other parts of Ethiopia. McCann's story, in essence, is one of a century-long erosion in the viability of rural households in Lasta: population growth, environmental stress, and transformations in Lasta's sociopolitical situation. His story properly starts in 1892 with the end of the Kefu Qan, following which recovery began, which was countered at much the same time by the onset of the demographic, political, social, and environmental changes, which undermined household viability.

McCann hypothesizes that since 1892, northern Wallo has seen a steady population increase, which has expressed itself locally in the territorial expansion of cultivation to higher elevations and downward to lower altitudes to occupy lands previously used by pastoral peoples. He argues that population pressure is evident in agricultural intensification in long-settled regions and in the emigration of Lasta farmers to many other parts of Ethiopia. The primary evidence is the testimony of Lasta informants, supplemented with information from informants in an analogous situation in the lowland areas below the escarpment of northern Shawa, and with aerial photographs taken in the neighboring Simien Mountains in 1954 and 1974. Also contributing to population growth were "the social institutions of marriage, concepts of family and inheritance"[17]

The transformation of Lasta's sociopolitical conditions was largely the product of its increasing involvement with the national state, which led to an overall increase in the amount of taxes Lasta farmers had to pay and to the conversion of tax and tribute obligations from a proportional to a fixed rate. "Fixed sums, as opposed to payments in percentage of harvest," McCann argues, "significantly increased risk of famine in poor production years."[18] The national state undermined the position of the local rulers in Lasta by cutting them out of customs collecting, preventing them from raiding neighboring lowlanders, and restricting their rights to redistribute tax revenues. These measures added to the burden on local farming households as their rulers set about compensating for their loss of income by introducing "a plethora of new taxes and assessments," levied mostly in fixed sums. Tax rates began to rise, as did demands for corvée labor: "Thus by the late 1920s and early 1930s, the fiscal demands on the peasant household had altered considerably from a few decades before. Not only had taxes, demands on labor, and extralegal drains on cash supply increased, but they had done so in ways that considerably increased the risks to the household economy."[19] McCann identifies drought, pests, and human and animal diseases as the principal environmental stresses with which Lasta farmers had to contend and which severally and in combination resulted in serious loss of life. Drought occurred most probably in 1905, 1914, and 1927–28; locusts invaded in 1905, 1917–18, 1927–28, and 1928–29; and infectious diseases appeared with heavy loss of life in 1906 (cholera and dysentery), 1917–18 (rinderpest), and 1918–19 (influenza, also "a fierce type of typhoid"). Famine, as such, appears explicitly in the sources only in 1914 and during the Italian occupation from 1935 to 1941.[20] McCann's vision of the Wallo landscape is somewhat Malthusian—a landscape pauperized by farming practices driven by population growth, which resulted in deforestation and an "advanced state of erosion." However, there would appear to have been little direct evidence for these processes in Lasta from 1900 to 1935, and McCann later rejected deforestation and erosion as part of a misleading narrative.[21] His ultimate conclusion is that "overall, the region's vulnerability to famine appears to come less from drought than from a shortage of productive resources—land, oxen and forage—some of which may be a result of lack of rain."[22] Chapter 3 argued that drought was the principal environmental trigger to famine in Wallo in both 1973–74 and 1984–85. Here McCann draws attention to other environmental factors contributing to farmers' vulnerability to famine, a subject to which Mesfin Wolde Mariam has also made an important contribution.

The World of the Wallo Farmer: Three Influential Studies

Mesfin Wolde Mariam, whom we met in chapter 2, went on to devote his research energies to deepening our understanding of the vulnerability of Ethiopian farmers to famine.[23] *Rural Vulnerability to Famine in Ethiopia 1958–1977*, published in 1984, covered the whole country and assessed famine vulnerability according to data gathered at the level of the *awraja*. At the time of the study there were 102 *awraja*. Mesfin explained famine vulnerability according to level of subsistence, agricultural commercialization, environmental quality, and degree of development. It was based primarily on documentary materials—published reports and government archives—but also informed by Mesfin's extensive firsthand experiences. Its conclusion was that famine was a problem affecting most parts of the country, only 7 of the 102 *awrajas* escaped the affliction between 1958 and 1977, and it was a function of subsistence agriculture, rather than commercial agriculture. The lower the level of subsistence, the greater the vulnerability to famine; conversely, the higher the level of development, the lower the vulnerability to famine.[24] A "system in which the majority of peasants are totally dependent on the physical environment and on their backward methods of production, and in which the socio-economic and political forces persist in incapacitating the productive potential of peasants by incessant oppression and exploitation is a condition for vulnerability to famine" (169).

Whereas *Rural Vulnerability* recognized that "the quality of the physical environment is very important. Famine is associated with deficiencies in the physical environment,"[25] it did not highlight particular environmental factors such as drought. In stressing how widely famine occurred, the book may have obscured the extent to which some parts of the country had been more vulnerable than others. To be sure, famine in 1984 was widespread; however, from 1972 to the present two parts of the country appear particularly vulnerable—the northeast highlands, running from south to north from northern Shawa through Wallo and Tigray to Eritrea; and eastern Ethiopia, from the border with Somalia westward toward Harar. Appearing as it did in the famine year of 1984, *Rural Vulnerability* was widely drawn on in subsequent studies. It focused attention properly on the plight of Ethiopian farmers. Mesfin directed his attention to this subject.

Suffering under God's Environment: A Vertical Study of the Predicament of Peasants in North-Central Ethiopia (1991) was funded by the Swiss Directorate for Development Cooperation and Humanitarian Aid and based on fieldwork in southern Wallo and northern Shawa carried out from 1986 to 1988.[26] The fieldwork involved questionnaires administered to more than 1,400 peasant households in 34 Warada and 285 localities and further involved measurements of altitude, field observations and interviews, and information from *Warada* offices. It presented its findings in terms of altitude, parsed into 200-meter intervals. The culturally determined agroecological categories of *dagga*, *wayna dagga*, *qolla*, and *werch* [(frost-prone uplands), grouped into seven zones, determined in the first instance by strict measurement, this time generally at 400-meter intervals.[27] Its findings covered a wide range of issues from perceptions of environmental change to demography, farming, livestock, trade, cash, and nutrition.

Mesfin's informants had interesting things to say about deforestation. Mesfin rightly notes that informant perceptions were bound to have been influenced by the environmental propaganda promoted by the agents of conservation and development. About 59 percent of informants believe that clearing the land for cultivation and settlement was a principal cause of deforestation; 2 percent blamed it on "burning"; the remaining 39 percent blamed both causes equally. As far as benefits of deforestation were concerned, 56 percent saw them in the new land made available for cultivation, 8 percent in the availability of increased grazing land, and 13 percent in getting rid of wild animals. As far as the problems brought by deforestation were concerned, 29 percent blamed it for bringing drought, whereas only 10 percent believed it would increase soil erosion; 22 percent named the shortage of firewood as the leading problem; and the remaining 39 percent blamed it for a combination of the already named factors. When these questions were posed in more open-ended form, the responses were rather different—62 percent of farmers believed that wildfire was the principal cause of deforestation as opposed to those who blamed it on the search for firewood (7 percent), the expansion of cultivation (5 percent), and for construction material (4 percent). Although recognizing some of the "impressive" achievements of the enclosure and reforestation program, Mesfin notes that these programs imposed very real costs on local farmers, starting with the people evicted. An "overwhelming majority" of farmers (83 percent) thought these programs should be confined to lands not already being used by the community.[28]

Population is another subject Mesfin explored with his informants. Households vary greatly in size, ranging from 1 to 10. The median size was 5.5, which is significantly higher than the averages reported by the Central Statistical Office for the Wallo *awrajas* of Qalu (4.4), Dessie (4.3), and Ambassel (4.2). I have no explanation for this discrepancy. Migration is a population-related subject, which Mesfin explored, unwilling, like so many other students of rural Ethiopia, to assume that Ethiopian farmers were unwilling or unable to move. Although 96 percent of his informants confirmed the standard picture in their disinterest in moving, a somewhat different reality was indicated by 22 percent of households, which experienced emigration firsthand. Divided more or less equally according to sex, about 6.5 percent of the population did emigrate. This was a somewhat youthful group, two-thirds of the emigrants being in their twenties or thirties. Family connections form an important part of the framework for emigration. Emigration is a considerably larger affair, two and a half times, than immigration, which affects only half the number of households (about 11 percent) (53–75).

In terms of self-perception, Mesfin's student-administered questionnaires found that almost three-quarters (73 percent) of his informants saw themselves as involved in both crop production and animal husbandry, as opposed to 27 percent who saw themselves as crop producers only. As far as the goal of their crop production is concerned, the survey found that an overwhelming 96 percent viewed it as subsistence rather than for the market, partially confirming one of Mesfin's major hypotheses, that "in Ethiopia peasant production is largely subsistence production and whatever enters the market is a consequence of coercion rather than a voluntary transaction" (84–85). This datum in itself says nothing about the forces leading to or inhibiting farmer market participation.

Mesfin rightly notes that accurate measurement of farm sizes is an impossible task. While informant perceptions are necessarily imprecise, they produced consistent results, such that the average farm size, across the peasant associations, which Mesfin studied, was 1.0 hectare. On average such a farm consisted of 2.3 fields. Surprisingly, a majority (51 percent) deemed that their farm was of an adequate size. As altitude increases, average farm size becomes smaller, but surprisingly again, a majority judged their farm to be adequate in size. Mesfin accounts for this, quite plausibly, by attributing it to a fear of resettlement, since the higher altitude zones were those most intensively subjected to resettlement (90–91).

The crops planted by Ethiopian farmers in the northeast were overwhelmingly cereals, followed by pulses and then oilseeds. Mesfin notes that eleven different kinds of cereals are grown, the most important of which are *tef*, barley, wheat, sorghum, and maize, along with five types of pulses and five types of oilseeds. *Tef* is under attack by agronomic experts as of comparatively low yield. These same authorities ignore its high cultural and apparently nutritional value and the fact that, as Mesfin notes, it is grown across a wide variety of farming conditions. In 1982–83 sorghum and maize accounted for 43.6 percent of total grain production, and barley and wheat for another 39 percent. This was a comparatively good year considering what was to come. In 1982–83 only 94.5 percent of farmers produced any crops at all. In 1983–84 this number fell to 84.3 percent and in 1984–85 to a catastrophic 59.3 percent. In lower altitude zones (below 1,800 meters) the number fell to 10–20 percent. By long-term standards, 1982–83 was a year of low production. All crops declined to 81 percent of their level in 1983–84; to 64 percent in 1984–85; and, in zones below 1800 meters, overall production fell to 10 percent of the 1982–83 level.

Ultimately, the way out of this impasse must lie in some degree of modernization. Artificial fertilizers would appear to be one of the most pertinent options. Polled on the use of fertilizers, only 24 percent of Mesfin's respondents were able to report having tried to use them. Attitudes appear rather complex: 40 percent of respondents felt that the main obstacle to the adoption of artificial fertilizers was simply their unavailability, whereas another 23 percent felt they were too expensive. Six percent believed they were simply not necessary, and 31 percent "did not know enough about fertilizers to want to use them." On the other hand, 75 percent of respondents claimed to use dung to fertilize their land, and only 5.3 percent claimed to practice crop rotation to maintain soil fertility (96–100). Meanwhile, productivity remains very low, such that Mesfin judges that "the peasants in this region live just below the level of bare subsistence" (104).

I explore Mesfin's findings on these issues more fully in chapter 5. Here the study's principal conclusions are of greatest interest. A central problem for Mesfin was the lack of rural transformation away from poverty toward more productivity and prosperity. This he attributed to three categories of factors—structural, institutional, and cultural. Several of these factors were discussed in chapter 3. The structural factor was created by the land reform and its implementation, which resulted in farms that were and still are too small and fragmented. The institutional factors included the constraints the land reform legislation placed on land accumulation and the selling of rural labor, excessive taxation and other levies, and the workings of the Agricultural Marketing Corporation (AMC).

In summarizing the institutional factors, Mesfin pointed to a thread that runs prominently through the various incarnations of the Ethiopian state, from the Kefu Qan to contemporary times, which is hence implicated in the three great famines considered in this book—the alienation of the ruled from their rulers, the chasm that separated the interests of the latter from the interests of the former. He observed, "The central problem in rural administration is the very lop-sided priorities of officials. Official priorities are not at all the priorities of the peasants" (187).

The cultural factors underlying the inability of rural Ethiopia to transform itself in the direction of prosperity and increased productivity, Mesfin identifies most clearly in his indictment of the market, an institution fundamental to the analyses of Sen and his commentators. The "peasants cling insecurely," he wrote, "to their subsistence goals of production,"[29] which he attributed to their deep commitment to the value of self-sufficiency and a distrust of markets.[30] Mesfin believed that the markets invariably worked to exploit farmers. There can be no doubt of farmer insecurity, farmer commitment to self-sufficiency, or the fraught relations farmers have with markets. Nonetheless, as we will see, farmers had solid ground for obstinacy in resisting many of the changes the state was trying to foist on them. They might distrust the market, but they also turned willingly to it.[31] However fraught and however modest even by Ethiopian national standards, participation in markets was important to Wallo's farmers. Mesfin's strictures did not close the discussion of the role of markets in rural life, as we will see when we review contributions by, most notably, Dessalegn Rahmato.[32]

Alemneh Dejene's *Environment, Famine, and Politics in Ethiopia* appeared in 1990.[33] It drew on Mesfin and McCann and helped the MacArthur-funded research team shape its field agenda. Alemneh carried out his research primarily in Wallo from October 1987 to April 1988. It was funded by the Energy and Environmental Policy Center at the John F. Kennedy School of Government, Harvard University, and the Rockefeller Foundation. Its goal was to identify "the forces that feed the interaction between environmental degradation and famine at the household level" and examine "the impact of government policies upon these forces" (4–5). The project gathered its data by means of a survey questionnaire and participant observation. It interviewed 230 randomly selected household heads from forty-six different peasant associations selected according to ecological zone, degree affected by famine, existence of land degradation, and extent to which the household composition was held by local officials to be representative of the *Warada* in which they were situated (6). (For a map of the *awraja* and Peasant Association, see figure 4.1.) Alemneh took the degradation narrative as established, and his research was designed more to illustrate it and grasp its underlying dynamics than to test its validity. He wrote of "the vicious cycle of environmental degradation, drought and famine" and attributed it to "the adverse impact of human activity on the environment." He held three activities as responsible: "overgrazing, overcultivation, and deforestation." He repeated the EHRS's conclusion "that 18 percent of the highlands will be bare rock by the year 2010 and 10 million people will not be able to produce food from the land."[34]

With his informants, Alemneh explored two centrally important topics: deforestation and erosion. *Environment, Famine, and Politics* accepted that deforestation played a major role in bringing about the collapse of peasant farm production into famine. Resting on a World Bank study, it set forth as coherent a version of the deforestation narrative as is available. Farmers were harvesting the forest faster than it could reproduce itself. They sold the timber to urban dwellers and increasingly relied on dung and crop residue for their fuel needs. With the tree cover removed, the use of dung and crop residue increased. Dung became the only source of fuel and a major source of cash for farmers, whose topsoil was disappearing. Erosion became dramatic. The process culminated in the disappearance of the soil's organic content. Further deforestation disrupted "the equilibrium that affects the

Figure 4.1. Alemneh's *awraja* and peasant association work sites in Wallo
Source: Dejene, *Environment, Famine, and Politics in Ethiopia*, 7, figure 1.2.

rainfall pattern" and brought about the collapse of farming, triggered by drought. Mass starvation ensued.[35] So the process started with deforestation, proceeded via drought, and culminated in erosion.

Alemneh relied on others for establishing the fact of deforestation, although he did remark, without reference to any historical baseline, that "the bare mountains and hills along the main road from Addis Ababa to Asmara testify to the extent of deforestation in Ethiopia." He accepted the idea that this process must have predated the twentieth century, but then returned to the assertion that "deforestation may have accelerated by as much as 40 percent in the last century." Asked to explain this, farmers attributed it to the increase of land under cultivation, increased demand for fuel and construction material, and expansion of livestock grazing.[36]

To Alemneh, soil erosion was the biggest environmental threat to the farmers of Wallo. He discovered that his informants did not share this view. Of 230 respondents, a respectable 26 percent reported that erosion was a "severe" to "very severe" problem. However, a strong majority of 73 percent reported only "minor" or "no problem." He believed these responses substantially underestimated the scale of the problem. His perception indicated the opposite; moreover, he thought the farmers had a plausible motive for underestimating the threat from soil erosion. The extent of soil erosion on a farm was a major criterion in selecting peasant association members for the reviled resettlement program. His conclusion, testimony of his informants to the opposite, was that "The Ethiopian government's assertion that most of the land in these peasant associations was unsuitable for agriculture, highly degraded, and should be under area closure to regenerate appears to have been valid."[37] This, in turn, in his view, validated the program of resettlement, "if implemented properly." This judgment proved highly controversial.

Dessalegn Rahmato's *Famine and Survival Strategies: A Case Study from Northeast Ethiopia* appeared in 1991, the same year as Mesfin's *Suffering under God's Environment*.[38] The study was commissioned by unnamed NGOs. It is based on information gathered at the level of the province (Wallo) and on field data and official records gathered in Ambassel, one of Wallo's twelve *awraja*, chosen as a microcosm of the whole and because Ambassel offered sites grievously affected by famine and some completely untouched sites. A small team of research assistants administered a questionnaire to 255 peasants between October 18 and November 5, 1986. The main goal of the study was "to examine the survival strategies of a peasantry faced with famine, and particularly at a time when external assistance is either nonexistent or of very limited reach" (15). It met its goal and provided an account of survival strategies of lasting value that were fuller and more nuanced than preceding studies.

Dessalegn began his account by stressing that Wallo responses to famine were characterized by cooperation and a community spirit, which previous studies had failed to discern, in spite of ongoing social inequality and a fundamental cultural division in the presence of Christianity and Islam. The Amharic language was universally spoken among the farming peoples of the highlands of Wallo and was one marker of a set of family and community values that overrode religious differences. Cooperation was conspicuous among those values and expressed itself in sharing assets and resources and caring for the sick and the infirm. Tesfaye Wolde Medhin, who studied local society in Warra Himano, an *awraja* in central Wallo to the west of Dessie, identified *wanfal*, reciprocity, or "relations of mutual support and cooperation," as a fundamental cultural value and a cornerstone of sociality.[39]

The possession of assets other than land was the basis for such social inequality as did exist. Most important among these assets were extra labor power, stemming from larger households and a full team of oxen. Additional assets, whose unequal distribution also contributed to social inequality, were livestock in general (cows, sheep, goats, equines), different kinds of movable property, and intangible personal qualities of energy, entrepreneurship, and initiative. A well-endowed farmer was publicly acknowledged as *bala gulbat* or *gulbat yallaw*, "one who has force."[40] Social inequality, in Dessalegn's view, did not express itself in classes but in strata, of which he perceived three—poor (or lower), middle (self-supporting), and upper, the last not to be confused with "rich." Although this is a useful description, it does obscure the fact that the distinction of one stratum from another is often indiscernible, that households are arrayed across a continuum from poor to upper, and that over decades one household might pass from one stratum to any of the others and back again.[41] The situation was highly labile.

Dessalegn describes Ethiopian highland sedentary agriculture as a peasant mode of production with the following four defining characteristics: (1) "The family is the unit of labour, of production and consumption." Production is directed to consumption, not accumulation. (2) "The market plays a supplementary role in peasant self-maintenance." The rural market is important for household resource management but does not determine land use practices or cropping plans. (3) "The stock of technical knowledge employed in production is the same for all households." (4) Periodic food crises are inevitable.[42]

His account of farming practices is nuanced and sympathetic but realistic. Twice he repeats the observation that "peasant agricultural practices are in the long-run environmentally harmful" (81). He argues that farmers are well aware of the need to reverse the processes that were sapping their soil of its nutrient value and that they have historically validated techniques which are effective, at least in the short run. With fallowing a practice of the past, the main farmer practice for rejuvenating the soil is crop rotation, with regular planting of beans and other legumes restoring nitrogen to the soil. Farmers do generally plow along contour lines, which helps block run-off. They build trenches around their plots to help drainage. Older, local practices of contour terracing were largely displaced by the more elaborate bunding schemes advocated by conservationists. To limit the most serious kind of erosion which occurs when the fields are bare, between final preparation for seeding and eventual germination, Dessalegn reports the use of rationing and mulching. Ratooning involves harvesting the plant in such a way as to leave enough of the plant above ground for its roots to retain hold on the soil. This technique postpones (but does not eliminate) the point at which the soil is left completely bare. Mulching involves leaving as much crop residue on the field after harvesting for as long as possible.

Markets Revisited

Markets receive considerable attention in *Famine and Survival Strategies*, and Dessalegn's remarks are a useful point of departure for reflecting on this important institution. For Dessalegn, markets play an essential yet supplementary role "in peasant self-maintenance" in rural Wallo. They are important tools for managing a household's resources. While exchange of commodities is the dominant purpose, markets are also extremely important as sources of information for farmers, who have no access to any other media. Most farmers patronize several markets within a range of two to three hours walking and for special purposes will travel a full day or two. In 1983–84, a time in which commerce in foodstuffs was highly skewed by the famine, a survey indicated that Wallo farmers marketed 20 percent of their sorghum and horse beans and 16 percent of their *tef* and lentils.[43] They marketed only 5 percent of the remainder of their crops. Throughout the mid-1980s in Wallo, *tef* was invariably the cereal of highest market value. Sorghum generally fetched a price only about 60 percent that of *tef*, and barley even less. The market allowed the exchange of products of higher value for products of lesser market value but greater caloric content. This was an exchange that Wallo farmers proved willing to make, not least, as we will see, when famine threatened.[44]

Dessalegn distinguishes between three levels of markets—local, area, and regional. Local markets may be found in *warada* capitals and other towns of comparable size. Area markets are larger and more diverse than local markets and are patronized by rural traders and merchants. Examples of area markets in our part of Wallo would be Waldiya, Hayq, and Bati. Dessie and Kombolcha are regional markets, dominated by merchants. They also provide access to commodities that households view as essential but do not produce themselves. Cereal production varies from one agro-ecological zone to another, and markets allow a household to consume a wider variety of cereals than it can produce. Spices, onions, coffee, and hops are all consumed well beyond the areas in which they are grown. Markets also allow for places endowed with surpluses to supply food to deficit places. Among the many limitations on the availability of free markets for the produce of Wallo farmers was the legal requirement that they meet the demands of the Agricultural Marketing Corporation (AMC), whose prices were always lower than those available through the markets.[45]

Table 4.2 compares AMC prices with those prevailing on the open markets of Ambassel and Qallu in June and September 1986. In June AMC prices were half or lower than those of the open markets; in September, the AMC prices having remained the same, the open markets reflected the impact of the new harvest. The AMC prices remained lower, twice as low in some cases, even though the free market prices were coming down.[46] The famine of 1984–85 induced the AMC to lessen its demands from an average of about 25 kg to 10–15 kg per household. But the great majority of households could not meet this demand and were forced to sell livestock or other possessions on the open market to buy the food demanded by the AMC.

As the famine crisis deepened, farmers became more dependent on the market in an attempt "desperately . . . to break out of the subsistence system, and plunge into the cash economy and the exchange system."[47] Hard cash, not barter, was the goal. Early on livestock was the asset farmers most relied on. Doing so eventually became counterproductive as cattle flooded

TABLE 4.2. AMC and free market prices of grain (free market in Ambassel and Qallu *awrajas*)

	AMC price (birr/quintal) *to peasants*[a]	*Free market prices* (birr/quintal)[b]	
		June 1986	*September 1986*
Sorghum (two grades)	27, 31	71.80	54.61
Maize	24	63.50	43.78
Tef (three grades)	39, 43, 50	119.63	100.73
Barley (three grades)	30, 30, 32	70.63	49.87
Wheat (three grades)	34, 35, 39	74.30	49.87
Horse beans	30	84.45	84.30
Field peas (two grades)	36, 40	108.00	88.85

Source: Dessalegn, *Famine and Survival Strategies*, 95, table 12.

Note: Free market prices for high-grade crops only.

[a] From AMC (June 1985): 63.

[b] From Records of the Red Cross, Dessie, October 1986. Average prices from markets in Ambassel and Qallu.

the market and prices collapsed. Other assets then found their way to the market—agricultural tools, clothing, family furnishings, jewelry, harnesses, hand weapons, firewood, and houses, which were stripped, some of the material going for firewood, some for building new homes. The demand for wage employment rose sharply and encouraged out-migration, not to local towns, but rather, as Dessalegn points out, primarily to the long-established sources of seasonal employment such as big plantations and large-scale rural construction projects.

In turning to the market, the famine-afflicted faced continually deteriorating terms of trade. The prices of cereals and pulses mostly peaked in May through July 1985, *tef* reached 204 *birr* per *quintal* in April and May (field peas soared to a high of 284 b/q in August 1985), while in October and November the price of an ox had fallen as low as 50 birr (one-fifth to one-third of its normal market value) and that of a cow to 35 birr.

Such food as there was in the Ambassel markets came from two main sources: food surplus *awrajas* in western Wallo, Warra Ilu, and western Dessie Zuriya, and from farmers in Ambassel who had escaped the worst effects of the drought. Dessalegn stresses that Wallo was afflicted with a relative, rather than absolute, scarcity of food for humans and feed for animals. A Red Cross study reported that in the neighboring *awrajas* of Ambassel and Qallu farmers suffered a loss of 73 percent in crop yield in 1984, but nonetheless, 66 percent of farms were cultivated. In Ambassel 25 percent of farming households had stored food enough to see them through from three to twelve months. Conversely, 75 percent of households had no harvest at all. In Qallu 10 percent of the polled had a harvest in 1984 adequate to feed their families for three to six months and 90 percent reported no harvest at all. By early 1985 international food relief, primarily in the form of wheat, was also making its presence felt.[48]

Back to History

In May 1996 Adhana Haile Adhana defended his dissertation for his PhD in history from Addis Ababa University—"History of Selected Famines in Peasant Societies in Tigray and Wallo, Ethiopia, 1941–1974." At 682 pages, it could not be said to have treated its subject in a trivial manner. To be sure, Tigray has suffered from neglect when compared with Wallo, in spite of Mesfin's early attempt in *Rural Vulnerability to Famine* to situate Tigray as the earliest province in northeast Ethiopia to suffer famine. Nevertheless, Adhana's dissertation does concentrate on Tigray and ends in 1974, with the conclusion of the first of the great famines. It covers three periods of famine: the Aradda Barre famines of the 1940s, which affected both provinces; the Tigray famine of 1958–59; and the Wallo famine of 1972–73, which also affected parts of northern Shawa and the southern Tigray districts of Rayya-Azebo and Wajjarat.[49]

Overall the 1940s in northeast Ethiopia appear to have been prosperous and productive in spite of the appearance of famine in 1936–37 and 1943. In the years around 1944, Wallo may have contributed 40.5–42.3 percent to the total national grain exports, which amounted to 30,000 tons.[50] Arrada Barre is a place near Marsa in Yajjuto where farmers of southeastern Tigray would go to buy grain. By extension the name came to cover a number of distinct famines, which afflicted Tigray and Wallo in the 1940s. The first year of these famines was 1936 EC (September 1943–September 1944). It was triggered by a plague of army worms in Addia Qayti. In 1946 drought afflicted a number of districts, causing people to migrate. The following year, 1947, saw generally good harvests; but in 1949 the *keremt* rains were inadequate and were followed by unusually heavy rains. Adverse climate hit again in the late 1950s and again from 1969 to 1973.

Famine in Tigray appears to have been limited in the 1940s. Only in Addia Qayti did it hit with any severity after 1946. Famine passed Wajjarat by in the 1940s. Agame and Endarta were affected only in their *wayna dagga* areas, in the first case by army worms and second by climate stress. Famine was rare in Maychew. There are indications that in 1947 throughout Tigray there was serious harvest loss as a result of drought, frost, and pests, but this was not universally confirmed by Adhana's oral informants. In 1944–45 (1937 EC) Makhoni and Dardar were badly hit by locusts, which destroyed many crops. In 1945 Rayya and Azébo were badly affected, first by the repression that accompanied the suppression of the Wayane Rebellion and then by an infestation of locusts. The years of most widespread famine were 1944–45 and 1945–46.[51] Famines in subsequent years would have been mostly *qolla* events.

In Wallo, information about the Arrada Barre famines is scattered. In general Wallo was in a more robust state than was Tigray. Yajju was a grain surplus area from the 1940s to 1966 vis-à-vis Tigray. Bati exported grain from Wallo to lowland Adal. In 1949 a severe drought hit Bati and put it in deficit. Various parts of Wallo were affected by army worms and/or drought, which brought famine, but Lasta, Wadla, Dalanta, and Dessie Zureya were surplus producers and suppliers of grain to neighboring, less favored districts. Again, as in Tigray, famine in Wallo was best characterized as predominantly a *qolla* phenomenon.[52]

Adhana devoted considerable thought to the causes of the famines. In Tigray he saw the suppression of the Wayane Rebellion in 1943, which was followed by infestations of army worms and locusts and by drought, precipitating famine.[53] He found the most defining characteristic of the Arrada Barra famines of Tigray as market dependence. In such a situation, the starving moved to the sources of grain, rather than the reverse. In Tigray the state had a meager role in creating or mitigating famine conditions. While the appropriate administrative elements advocated relief, the state itself saw the famines as "natural events." In Tigray there was no tax burden on which farmer impoverishment might be blamed; famine struck districts that, up until the early 1950s, had been viewed as prosperous; there was no land shortage and no land degradation.[54]

By contrast, in Wallo in the 1940s the administration effectively blocked the build-up of household food reserves and constrained the holding of adequate livestock assets. Unlike in Tigray, the administration of Wallo may well have laid the groundwork for the famines of the 1940s, which, however, were "more natural than social in their aetiology."[55]

In 1958 Tigray was struck by a province-wide famine that lasted well into 1959. Although the famine was uneven in its intensity, perhaps the hardest-hit district was Agamé, where most of the farmers migrated to neighboring territories, although in terms of absolute numbers the Adwa-Aksum area was struck more severely. Adhana is inclined to see this famine as societal, natural, and political in its roots. The societal dimension was a "function of peasanthood possibly marked by relative population expansion and therefore inevitable." Political origins lay in "stunting long-term political processes" and in a monarchical political system "which was congenitally flawed and insensitive as regards peasant economic and social well-being."[56] Up until 1958, politics in Tigray were a tussle between imperial officials, who tried to increase tax revenues, and local farmers, who tried to reduce them (367). Finally, drought and locust infestations were the natural forces at work undermining farmers' livelihoods, leading to the failure of the main harvest in 1958, which plunged the province into famine. By the summer of 1959, too late to prevent the loss of life, 9,000 tons of relief had reached Tigray. Relief food and a good harvest in 1959 brought this famine to an end.

Weighing the relative impact of drought vis-à-vis locusts, Adhana opted decisively for the latter, "locust invasion must be considered *the* cause of the Tigray famine," in the face of examples of locust invasions, which did not bring famine, and in the face of examples of famines, which had no association with locusts. The main locust attacks occurred between September and November 1958, and destroyed 167,000 metric tons of cereal grains. He estimated that of Tigray's 1 million residents in 1958, half lived in famine-affected areas and 130–150,000 lived under severe famine. He rejected as too high Mesfin's estimate that 408,000 were victims of this famine. The famine victims flocked to the towns.[57]

Death and emigration were the main social expressions of the famines. Adhana rejected any suggestion that articulation with capitalism was at work: "the society in 1958–59 was perfectly peasant rural as in the forties" (411). On the other hand, population growth may have reached its limits under the peasant mode of production. The burden of taxation had become unbearable by the late 1940s. Taxation did not pauperize peasants but "bled off assets which might have mitigated the impact of the famine" (418).

The Wallo famine of 1972–73 was similar in its origins to the Tigray famine of 1958–59, but owes its greater severity, in Adhana's view, to the extractive behavior of the state vis-à-vis local farming communities (578–79). Adhana understands the bulk of Wallo's farmers to have been under one or the other of two forms of land tenure, both of which allowed for farmer control of production. Production was for use value and not of commodities. Adhana also believed that all levels of Wallo's farmers were equally affected, although somewhat contradictorily, he also held that some households were well off enough to get by with their own reserves. Population increase in Wallo had led to a shortage of farm land, which led to an "ever greater dependence on nature" rather than on new social forms or primitive capital accumulation (538–42, 579).

Adhana contends that Wallo farmers bore a heavier tax burden than did the farmers of Tigray. Although it is true that from 1942 on the national government introduced a number of new taxes that bore heavily on farmers, most of these taxes ought to have been felt equally in Tigray as in Wallo. The measures included a root-and-branch reorganization of provincial government in August 1942, new regulations for the administration of church lands in 1942, new regulations for the organization of customs and export duties in 1943, a land tax in 1944, an education tax in 1947, a health tax in 1959, and an agricultural income tax in 1967.[58] Some of these measures provided for the elimination of the *asrat* or tithe, but in fact, many farmers were still paying this tax as late as 1974. As a result, when famine broke out, many were burdened both with a heavy load of taxation and, particularly in Wallo, an incompetent local government, rife with corruption.

In summary, Adhana described imperial Ethiopia as "an incubator of killer famines" (597). He went on to say that nature caused the famines in Tigray and Wallo, with approving reference to Michael Watts's dictum in *Silent Violence*, to the effect that the decisive causes of famine are always the "social referents" of the natural event: "The nature of the historically determined society determines whether or not adverse natural events necessarily culminate in famines" (600–608). The historically determined societies of Tigray and Wallo were peasant societies. As such, their productive capacity was essentially static. Their inability either to generate transformative change from within or creatively to respond to externally generated change.

Stasis?

In describing Ethiopian farming communities as essentially "static," Adhana makes explicit an assumption that runs through much of the literature. It is implicit all through the EHRS and would seem to underlie Mesfin's cultural obstacle to rural transformation. There are studies that find Ethiopian farming communities adaptable. Particularly nuanced is the approach of Alex Winter-Nelson, an economist and member of the MacArthur research group. Winter-Nelson found that some of households in rural northeast Ethiopia—in Tahuladare *warada* explicitly—were capable of generating modest surpluses and would be viable even under the devastating conditions of 1984. These households formed a majority, and there was a minority that were marginal under any foreseeable circumstances. What undermined the capacity of the majority of households was how the Ethiopian government distorted commodity markets.[59] Under the Darg, those households suffered from an inflated national currency, heavy direct and indirect taxes, and commodity prices that were aggressively manipulated by the government to favor urban consumers.

Winter-Nelson's conclusions are supported by a survey of the food economy of northeast Ethiopia undertaken by Holt and Lawrence and sponsored by Save the Children Fund UK. Holt and Lawrence documented the tightness of margins under which the region's farmers worked. Nonetheless, they found that to push the farmers into famine took several years of drought and least two consecutive badly failed harvests: "The conclusion we have been tempted to draw from this is that these poor, rural communities are remarkably resilient, and are able to withstand what equates to the loss of the best part of one year's income as well as of much capital through the death of livestock."[60]

A penetrating study of Ethiopia's historic, plow-based agrarian economy is by Berhanu Abegaz, who finds it in persistent stasis, which he attributes to "endemic extractive contests between a predominantly landed peasantry and a titled, prebendary overlord class. The latter's strategy of political accumulation inevitably engendered immiserization of overlord and peasant alike by privileging diversion over production. The surplus was then dissipated on unproductive consumption, national defence and internecine strife. Lacking a strong state to mitigate predation and political instability, the Ethiopian peasant rationally 'chose' to be efficiently, albeit self-sufficiently, poor."[61] Berhanu's account helps explain the general functioning of Ethiopia's historic agrarian economy. On the one hand, its farmers have proved to be creative, industrious, and adaptable—not least through their domestication and indigenization of a plethora of crops. On the other hand, in recent centuries they have made no systemic, transformative breakthroughs. Both Berhanu's account and Adhana's Marxian-rooted one seem to some degree overdetermined. They are systemic accounts, and neither provides for the openness to change or the capacity for innovation some of us have found in the life histories of rural Ethiopian informants, which I explore at greater length in chapter 6.[62]

Population

Unbridled population growth has been much remarked on as a contributor to environmental degradation and famine vulnerability. Markos Ezra produced the first extended treatment of the subject, one with special reference to northeast Ethiopia.[63] His working hypothesis is that "the rapid ecological degradation is apparently manifesting itself to the peasants in various forms. The most obvious manifestations include drought (failure of rains), declining sizes of landholdings or landlessness, and an accelerated forest resources depletion the result of which has been a persistent decline in food grain and livestock production as well as a recurrence of famine" (47–48). He believes that the ecological crises are stimulating changes in the demographic and socioeconomic lives of peasants, which in the near future will lead to demographic transition. This will include "changes in patterns of age at marriage, attitudinal changes towards large number of children, ideational changes on desired number of children, demand for family planning services and tendency and desire to look for non-farm employment."

In the context of northeast Ethiopia, Markos sees ecological degradation as manifesting itself in recurring drought and famine, irregular and unpredictable rainfall, land scarcity, declining land productivity, and lack of forest resources, especially of fuel wood.[64]

Markos's study used structured questionnaires to glean information from individual households, from women who had ever been married, and from community leaders. He conducted the survey over six months starting in mid-October 1994 in eight different *waradas*—two from north Shawa; two from Wallo (of which one was Tahuladare); and four from Tigray. These sites were chosen on the recommendation of the Ministry of Agriculture and the RRC as having been unusually severely affected by drought and famine over the previous two decades. He selected forty different peasant associations (known in Tigray as *tabia*) from the *waradas*, and from each peasant association fifty households and fifty ever-married women. Altogether 2,000 households and 2,000 women who had been married at one point were surveyed by assistants; Markos himself interviewed community elders and government officials at the *warada* level. He found these interviews to be most informative when community members were interviewed separately from government people.

Markos also used government documentation on population, land and forest resources, rainfall, agricultural production, and national records of drought and famines, but as he remarked, secondary data on these issues are scant. Different authors give widely different estimates of Ethiopia's cultivable land resources, but "Even more problematic is the measurement of the magnitude of ecological degradation . . . land and forest resources are often taken as main indicators of degradation. But measurement of the extent of land and forest degradation is extremely a problem" (56). Problematic as they are, Markos uses "an inventory of various sources of secondary data" to "analyze the macro-context or the background situation." In the end he is forced to rely heavily on the findings of the EHRS. With respect to forest resources, in addition to the EHRS, he gives prominence to a 1990 study sponsored by the International Union for the Conservation of Nature. This report repeats the discredited figures—in the early nineteenth century, 87 percent of the highlands were covered "with natural forest vegetation," which by 1900 declined to 40 percent of previous and by

1955 down to 5 percent. He notes that an Ethiopian Forest Action Plan exists. He also remarks, almost as an aside, "observation during the fieldwork has revealed that . . . forest resources are not frequented by many peasants except in Wello, where the habit of planting trees for economic benefits has been a norm since the last 10–13 years" (77). Tree planting by farmers in Wallo is discussed at some length in chapter 6, where it is traced to the 1930s and used as a prime example of farmer initiative, adaptability, and creative landscape management.

Markos's findings on farmers' perceptions of the standard degradation narrative are interesting. He approached the question by asking about land tenure, land fragmentation/distribution, scarcity, and productivity. His informants, both gainers and losers in the process of redistribution, see redistribution as problematic because of the very idea that anyone should lose land is abhorrent, and because it discourages investment in the land. Farmers associate redistribution with the fragmentation of holdings, which, they hold, also discourages investment. Asked if their present holdings were adequate for their needs, Markos's informants' responses diverged significantly from those of Mesfin's informants—most responded negatively, with little variation across peasant associations or *waradas*. I think we have to assume that calculations here would include their perceived implications for resettlement, with farmers in all circumstances seeking to avoid it. This would suggest that their anxieties about being resettled were at a low ebb. About 90 percent of informants said they felt that land was becoming scarce. Opinions were unaffected by biological age. Grazing land is a problem for those who own cattle; about 76 percent of households see lack of grazing land as a major problem. Grazing lands used by themselves and their parents in earlier times had been enclosed.[65]

To explore farmer awareness of the degradation of forest resources, Markos asked questions about distances traveled to fetch wood and water in the present compared with ten years previously. About 50 percent of his respondents reported that ten years ago it had taken them less than thirty minutes to fetch fuel wood, but today only 31.7 percent so reported. Water apparently shows less change, which Markos attributes to the increased number of wells available. Tree planting was one response to declining availability of fuel wood and occurred under three guises—state forests, community forests, and private plantations. The respondents made it clear that the state and community forests were of no value to them, and presumably some of them were among those farmers who destroyed so much of the government's conservation tree planting in 1990–91. Markos inquired about the extent (if at all) to which questions of environmental degradation were discussed within the family. Just under three-quarters of households reported family discussions of "the size and quality of landholding" and of "the lack and shortage of fuel wood," which indicated the level of their environmental awareness.[66]

Markos established that just over 90 percent of farmers believed that their lands were declining in productivity (table 7.13, 220). To find out what farmers thought about the causes of this decline, household heads were presented with a list of generally discussed causes and asked if they agreed or disagreed that these were causes of decline in productivity: 93.1 percent agreed that "poor rains" were a cause; 86.6 percent agreed that this was also the case with recurrent drought; 79.7 percent blamed the "aging of land"; 63.2 percent blamed the smallness of farm plots; and 55.7 percent agreed that the inability to fallow exacerbated things. Other potential causes—lack of fertilizer, overgrazing, and shortage of labor—received less than 50 percent endorsement. The only thing that calls for comment here is the relevance of the size of farm plots, because drought is widely recognized and the aging of land and lack of fallow are also intimately linked. A decline in overall size of farm plots leading to a decline in overall production is to be expected, but not that it would lead to a decline in the productivity of individual farm plots.

Markos's elderly informants felt strongly about deforestation and the visible disappearance of stands of trees across the landscape, which they claimed was a development of the past twenty years: "It was between 1970s and now that these forests were destroyed" (125). They believed the principal causes of deforestation to be clearing land for construction, settlement, and cultivation; charcoal and firewood production; and the recurrence of drought over the previous three decades. The consequences have been a shortage of firewood and construction material; soil erosion; and "reduction of rainfall and its consequent drought" (125), a cause-and-effect relationship taught to them by development agents (126).

Given that farmers recognize the principal problems they faced, Markos queried the failure of twenty years of development projects pursued by the government and NGOs. Most of these were funded through a food for work program. He feels that the truly underrated factor is the farmers' resistance. One example is their failure to adopt irrigation, even when, as in Hayq, the government invested heavily in infrastructure (the ruins of which were still visible when I was carrying out my research in 1997). Widespread was the practice of farmers sabotaging their work on food for work projects to keep the work going. However underrated farmer resistance to change has been as a contributor to the failure of the development plans of the 1980s, Markos does agree that the principal cause of failure was government policy, with its litany of misconceptions and misdirections as already recited by Winter-Nelson and others.

Markos argues that the problem with the EHRS is that it "focused on ecological rehabilitation without due attention to increasing productivity levels of the farming population," and failed to realize that conservation could only be sustainable if local people saw immediate benefits in increased production (133). He agreed with other researchers, notably Yeraswork (1998), that continuing insecurity in property ownership was a fundamental challenge to on-the-farm conservation activities. His informants, on the other hand, did not fully share this view. Just over 70 percent favored periodic redistributions of land in the interests of equity. Fully 75 percent of farmers were opposed to private ownership of land. Although one might wonder how fully farmers were aware of the issues at stake here, in fact consciousness of the issues had been much heightened by the discussions of the draft 1995 constitution, which had only recently taken place in all peasant associations and contained substantial options for property holding.[67]

Finally, on the question of whether they were in favor of re-settlement to places where land is abundant, only 9.5 percent of respondents indicated that they were, whereas over 90 percent said they were not (138–39). In spite of this widespread reluctance, resettlement had already started in 1976. Between 1976 and 1979, some 48,000 families from northern Ethiopia resettled in the southwest of the country. In 1984 resettled population totaled about 200,000, until November, when the government decided to move nearly 600,000 farm families from Eritrea, Tigray, and Wallo. Unpublished RRC reports indicate that between 1980 and 1990 the government resettled 343,000 households (roughly 1,715,000 people) in the western and south western parts of the country (253).

One of Markos's central concerns was to find out what (if any) demographic response there had been to degradation and famine. First he established that the 1994 census preliminary results showed slower population growth. The Central Statistical Office said that between 1984 and 1994 that national population growth rate was 2.2 percent for rural and 4.1 percent for urban areas, which combined were less than the projected 2.96 percent.[68] His conclusion was that particularly in the ecologically stressed areas, fertility had begun to decline. Three observations led to this conclusion. First, the age structures resting on data from 1984 and 1994 consistently show a decline in the proportion of the age group zero to four years. Second, the Central Statistical Office declared that between 1984 and 1994 rural population grew at an annual rate of 2.3 percent. Third, estimates of fertility parameters for Amhara and Tigray seem to indicate decline (298–99).

Markos summarized his findings. He concluded that northern Ethiopia was suffering from various kinds of degradation: erosion, overcultivation, and fragmentation. Average landholding per capita for the region is 0.17 hectares, whereas for Wallo it is 0.10 hectares—insufficient with existing technology for sustaining livelihoods. He believed that awareness of degradation problems was higher than ever in government circles. Farmers were more than aware of the threats to their livelihoods posed by soil erosion, forest depletion, and fragmentation of farm plots. However, many farmers believed that the standard antidegradation measures, such as hillside enclosures, ran against their interests by reducing land available for farming and grazing. But they supported some of the conservation activities, because the food for work programs provided them with vital supplementary income. Markos found that farmers found population to be a problem and that attitudes toward the ideal number of children were shifting downward, especially in peasant associations where land is scarce, such as Tahuladare. Farmers attributed declining field productivity to the fields having become "old," and to the land becoming smaller.[69] Markos addressed the question of food insecurity and argued that "food deficit in drought prone areas will persist in so far as modern farming techniques compatible with the problem of ecological degradation are not instituted."

Markos's central concern about whether there had been changes in fertility behavior in response to environmental degradation led him to research changes in age at first marriage, singulate mean age at marriage, number of times married and/or never married, and the proportion of divorced and widowed. He found delay in the age at first marriage of both men and women, as well as an increased rate and frequency of entry and exit into and from marriage. Fifty percent of his 1987 women informants reported having been married more than once, suggesting that there is declining stability of marriages; however, uncertainty about any long-term trends in divorce makes this speculative.[70] As we have already seen, Markos found a convergence of evidence to support the belief that fertility has begun to decline. He questions whether this is a short-term or long-term phenomenon.

Twenty Years to Nowhere?

Yeraswork Admassie, like Markos Ezra, contributed to the EHRS.[71] His major follow-up study was *Twenty Years to Nowhere: Property Rights, Land Management and Conservation in Ethiopia* (2000). Yeraswork carried out fieldwork in four trips, two to Wallo (January 27–February 16 and September 23–October 14, 1992) and two to north Shawa (October 25–31 and November 18–23, 1992) (see figure 4.2). He was accompanied by a number of field assistants, who administered his structured household questionnaire, while he conducted focus groups and carried out in-depth interviews. Wallo was his primary target, as it had been for the national soil conservation and afforestation program. North Shawa was secondary. He chose three catchments in Wallo where the program had operated longest: Alawuha in Yajju, Upper Borana in Dessie-Zuria, and Upper Mille in Ambassel.[72] Within each catchment, his strategy was to choose two to three peasant associations, one close to a national road and/or major urban center, one quite remote from these amenities, and one intermediate. This did not prove possible everywhere. The project's original goal had been to interview not less than 5 percent or fewer than sixty of the household heads who had been cultivating their own plots for not less than five years. This proved impossible in some instances. Peasant association records were in a shambles—and in some cases even the number of total residents or members had been lost. This meant that a truly random sampling was impossible.

The combined national and international effort for soil and water conservation originated in the crisis situation of 1972–73. With the crisis relieved, some agencies shifted to environmental rehabilitation work, using the surplus grain still available through food for work programs as inducement for farmer participation. The Ethiopian Ministry of Agriculture and Environmental Protection and Development initiated a number of soil and afforestation projects on its own with farmers mobilized through the peasant associations. Two units within the ministry—the Soil and Water Conservation Department and the Forestry and Wildlife Conservation and Development Authority—took charge of these projects, and their achievements were impressive in scope, as already noted in chapter 3. Yet

Figure 4.2. Yeraswork's study sites
Source: Yeraswork, *Twenty Years to Nowhere*, 47.

the emphasis was entirely on a speedy response to a crisis with little attention paid to long-term impact or sustainability.

The UN World Food Programme (WFP) "Project Ethiopia 2488—Rehabilitation of Forest, Grazing and Agricultural Lands" was the largest actor in the national conservation and reforestation program. The United Nations supported this food for work program. It was supported by more than 720,000 metric tons of wheat, 25,000 metric tons of vegetable oil, and "a wide range of non-food items." The total amount committed by the WFP over the fifteen-year period from 1975 to 1990, was about US$250 million. Its basic strategy was to operate at the catchment level, a level that might embrace multiple peasant associations. By the mid-1980s it operated in 117 river catchments. By 1987 the program had treated about 25 percent of the project's total area of 2.6 million hectares. It afforested 180,000 hectares and put in place soil conservation structures or revegetated 460,000 hectares.[73] The location of all bunding and tree planting was decided at the higher level and the benefits (or lack thereof) to any particular farm field were of no concern to the project's designers and managers.[74]

Food for work projects provided the inducements used by a host of other agencies to carry out their development and rehabilitation activities—the Ethiopian Red Cross Society, the Swedish International Development Authority, Deutsche Gesellschaft für Technische Zusammenheit, and the United Nations Sahelian Office (UNSO). The Ethiopian Red Cross Society was the principal sponsor of the Upper Mille and Cheleka Catchment Disaster Prevention Program, an integrated conservation-based rural development project heavily engaged in tree planting and soil conservation. One of its main claims to fame was that it tried to introduce a participatory approach. The UNSO set up a couple of large-scale fuel wood plantations, one of 3,600 hectares near Dabra Berhan, and one of 4,200 hectares near Dessie.

The program's short-term achievements were notable. Many different catchments saw conservation interventions, although the total area covered was only 2 percent of the degraded land at the national level. Expansion was restricted by the limited resources of the food for work program, and Project Ethiopia 2488 proved unable to extricate itself from the initiatives it had started. Doubts grew about long-term viability. Between 5 and 25 percent of soil conservation structures were damaged within the year of their construction, and as this problem grew, the project became increasingly focused on protecting its assets. It became clear that the program was bringing few tangible benefits to farmers. Although the amount of fodder, firewood, and construction materials grew over the years, these increases were not available to farmers. Yeraswork's central research problem was "why soil-conservation and afforestation technologies and practices induced by external agents through a project fail to be self-sustaining." He refined his research problem as "the long-term individual and collective adoption and sustainability of three objective groupings of project activities: (1) on-farm soil conservation measures (including private tree planting) (2) off-farm afforestation, and (3) off-farm soil conservation measures."[75] To this problem he added an interest in determining what role land tenure, current and from the immediate past, played in the decisions farmers made about adopting new technology or participating in the collective work of environmental rehabilitation.

Yeraswork had planned his project during the last years of the Darg. By the time he arrived in the field in 1992, the program he had hoped to study had collapsed. His task changed to explaining this collapse and assessing any ongoing influence the project may have had. His summary conclusion is that it is "as if written on water."[76] The soil conservation measures had either been abandoned by the authorities or intentionally destroyed by farmers. In many areas farmers had gradually plowed under the bunds and other soil conservation measures during the last two years of the Darg; in some areas, this happened more precipitously with the Darg's fall. The farmer in possession of the plot did away with bunding. Yeraswork's team heard no evidence of vigilante destruction in neighboring farms.

The peasant association hillside plantations suffered similar fates, although obviously not at the hands of those who had individual claims on them, since few (if any) had. While privately owned trees escaped virtually unscathed, the plantations were ruthlessly looted by the more aggressive or avaricious members of the local peasant associations. Again, there does not appear to have been organized vigilante activity. Historically established natural woodlands were treated more gently, sometimes emerging untouched. The famous "One Million Tree" plantations of the Upper Mille and Cheleka Catchments Disaster Prevention Program were recklessly destroyed, their trees chopped down in such a manner as to preclude regeneration. In some places, according to a soil and water conservation expert, the trees were actually uprooted.

With the establishment of a new government in Addis Ababa, following the ouster of the Darg, the peasant associations were reorganized to a degree and given control over the plantations, which hitherto had been controlled by the Natural Resources Conservation Department. This led to a fresh round of deforestation, carried out by the farmers themselves: "Everybody participated, and there was competition to get the most. Whole households—those that have three members, including the women—all went into the plantation. In competing for possession of the wood, people behaved as though they were 'unrestrained dogs.'"[77] Only a few peasant associations managed to temper the illegal logging. Yeraswork found no evidence to suggest that the deforestation had any Robin Hood/redistributive dimensions.

One of his more distressing conclusions was that such technical knowledge relating to reforestation and soil conservation as had been disseminated either had not survived or had not been disseminated in the first place. The Ministry of Agriculture, Environmental Protection, and Development, which had been in charge of the conservation projects for some time, preferred to work with "small tightly knit groups that jealously clung to their monopoly of technical expertise" (85). This narrow layer of expertise was continually eroding.

Almost unique among researchers in this area, Yeraswork recognized the existence of tree planting as a significant farmer activity. His household survey confirmed that tree planting was widely practiced, but his presentation rather minimized its significance. He started out with the negative observation that nearly a third (31.5 percent) of surveyed households said that neither they nor their parents had planted any trees before the land reform of 1975. Only then do we learn that the remaining two-thirds of the households had planted trees before 1975. One-third (32.6 percent) had successfully grown up to 50 trees in all; another 17.5 percent had planted between 51 and 150 trees; 8.5 percent had planted 151 to 350 trees; and 10 percent said they or their parents had planted more than 350 trees.[78] Tree planting was strongly related to freehold or usufructory landholding, and tenants were unlikely to engage in it.

Tables 4.3 and 4.4 show that tree planting was an activity of consequence before the land reform of 1975, that across the eleven peasant associations individual farmers planted an average of almost 78 trees each, just under 60 percent (58.8 percent) of these planters were owner-cultivators, just under a quarter (24.0 percent) were tenants, and the remaining 17.2 percent were usufruct holders. In general the more active tree-planting peasant associations were closer to a major urban center or a national highway.

Table 4.3. Mean number of trees planted before the agrarian reform and percentage distribution of respondent by pre–agrarian reform user tenure status by peasant association

		Respondents' pre–agrarian reform user status		
Peasant association	*Mean tree holding*	*% Owner cultivator*	*% Usufruct holder*	*% Tenant*
Wuraf	33.2	35.4	16.9	47.7
Wayneye	140.4	74.0	0	26.0
Machare	55.6	40.8	19.7	39.4
Bulbulo	89.6	59.0	19.7	21.3
Amamo	80.0	68.9	18.0	13.1
Qalam	131.5	57.1	25.4	17.5
Tossa	82.5	68.9	26.2	4.9
Hara	95.8	98.2	1.8	0
Yadi	22.4	4.3	61.7	34.0
Andit-Tid	61.7	66.7	0	33.3
Laimush	68.8	73.3	0	26.7
Overall	77.8	58.8	17.2	24.0

Source: Yeraswork, *Twenty Years to Nowhere*, 94, table 2.

TABLE 4.4. Mean number of trees planted by respondents before 1975, by pre-1975 own/parent and own land user tenure status

	For whom it was:			
	Own/parent's status		*Own status*	
	Mean # tree holding	*# of Cases*	*Mean tree holding*	*# of Cases*
Owner-cultivator	93.7	422	87.1	369
Usufruct holder	75.3	136	82.0	104
Tenant	49.8	187	57.6	150
Multiple status	38.2	25	47.7	645
Overall	78.0	770	78.1	645

Source: Yeraswork, *Twenty Years to Nowhere*, 97, table 3.

Such was the case with Qalam, near Dessie, where privately run plantations sprang up to serve the urban market for construction material and fuel. The natural vegetation at the highest inhabited areas is grass rather than trees, and as a consequence, the highest of the peasant associations generally evince a low level of tree planting. Yadi would be a case in point. Informants offered a number of responses when queried about obstacles to tree planting. Yeraswork determined that there were two of significance: the availability of wood from natural forests and prerevolutionary landholding status, with owner-cultivators the most likely to plant trees. He found no evidence that any of the officials or agents of Ethiopia 2488 were interested in encouraging tree planting by individual farmers.[79] The trees planted by these informants were overwhelmingly eucalyptus, but Yeraswork noted that they also planted other perennials—*gesho* (a hopping agent), sugarcane, and coffee, "albeit on a small scale."[80]

Yeraswork offers several plausible explanations for the resistance of Ethiopian farmers to adopting bunding and reforestation. At bottom, he argues, resistance stemmed from the fundamental insecurity the 1975 land reform brought to the holding of agricultural land. First, it was rarely out of the farmer's mind that the "owner" of the land was now the government, and the government could and did expropriate farmers for a host of reasons—soil conservation and reforestation not least among them. Insecurity was reinforced in a host of ways—most obviously through the periodic redistribution of land carried out by the Darg and the succeeding EPRDM government. The frequency and intensity of redistribution varied considerably from peasant association to peasant association, but all farmers experienced at least one, and many lived through several. Some redistributions violated fundamental principles of equity. Any number of farmers lost land in redistribution after they had prepared it for planting. In 1984–95 one farmer lost 3,000 eucalyptus trees and 4,000 sugarcane plants without compensation. He had to pay B600 to buy construction wood for his new house, while the farmer who had benefited from this redistribution got B700 by selling wood on the market after having built houses for himself and his four children. Yeraswork makes his case that the insecurity of their land holdings was a major factor discouraging farmers from adopting bunding. What I find remarkable in this situation is not the numbers of farmers deterred from conservation measures or engaged in destructive behavior but the numbers of farmers who continued to plant trees in spite of so many obstacles.[81]

Yeraswork's account of the failure of the grand plans for reclaiming the Ethiopian highlands is echoed in Dessalegn's essay "Littering the Landscape."[82] Together they constitute a major indictment of plans for radical landscape change and mark a suitable point for redirecting our attention to an assessment of the nature and extent of landscape change and the role of Ethiopian farmers in generating that change.

5

Parameters of Landscape Change in Wallo

Landscape change was central to attempts to explain famine vulnerability in twentieth-century Ethiopia. The notions of deforestation and desertification entailed visible impoverishment of landscapes. Both implied a loss of vegetation, woody vegetation in particular, which was attributed to the expansion of farming driven by population growth. In their observations of the highland landscapes, commentators saw clear evidence that these processes were at work. Their observations rested on flimsy foundations. No baseline had been established against which to measure change. The history of Wallo landscapes was unknown to them. Wallo was not unique in this respect, for historical landscapes in Ethiopia have been very little studied. Subsequent research, in which images from photographs and remote sensing have played an important role, has advanced our knowledge, particularly of the twentieth century. Careful examination of written materials has established a baseline in the mid-nineteenth century. It tells us that Ethiopian landscape change is more subtle, more complex, and less dramatic than the initial accounts allowed.[1]

Ethiopian Landscape Change

Serious source-based studies of landscape change in Ethiopia began in 1978 with Stähli's study of settlement and land-use in and around Simien National Park between 1954 and 1975. The breadth of its source materials and the rigor with which they were interpreted set a high standard for studies that followed, while its larger interpretative framework was problematic.[2] The study was the first serious attempt to use photographic evidence—terrestrial photographs taken in 1954 and aerial photographs taken in 1964—which it supplemented with a field survey in 1975 and the accounts of European travelers from the late eighteenth century to the beginning of the twentieth century. Stähli was concerned with the relationship between environmental change and population growth. The study was sensitive to the potential differences between "highland" and "lowland" agro-ecological zones. Using changes in the number of farm buildings to measure population growth, Stähli confirmed national figures that had been extrapolated from sample surveys. He concluded that the upper limit of farmland had climbed about 100 meters to 3,750 meters above sea level (34) and that he could discern the impact of population growth in the clearance of woody vegetation and the farming of increasingly steep slopes.

Stähli's broader interpretative framework was conservationist, setting its highest value on the preservation of local flora and fauna, and was alarmist in tone. He buttressed his observation about woody vegetation with reference to the now-discredited claim that "85 percent of Ethiopia's original forests have been destroyed," leaving only 5 percent of the country covered by trees.[3] This he attributed to "the age-old agricultural tradition of clearing the land by burning and the lack of understanding on the part of the native farmers and the local public authorities of the ecological problems."[4] He concluded that the situation "will lead inevitably to a catastrophe." He foresaw no capacity for adaptation in "traditional agriculture" and advocated external intervention to improve both agriculture and forestry.[5]

More than twelve years elapsed before the challenge Stähli laid down was taken up in the 1990s, by which time there were studies of change, revisionist of prevalent environmentalist thinking, in a number of other African landscapes, which buttressed studies of Ethiopian and Eritrean landscape change, which started to appear at the same time. African landscape studies in the 1990s drew on a rich trove of sources and deployed new methodologies. In the case of Ethiopia, a comprehensive national aerial photograph in 1986 established a new benchmark for change when compared with an earlier aerial photograph taken between the mid-1950s and mid-1960s. Supplemented by evidence from a wider array of terrestrial photographs and a more thorough reading of written descriptions of the landscape, this material was the basis for a body of literature that nuanced our understanding of environmental change in Ethiopia.

Two particularly influential African studies were by Tiffen, Mortimore, and Gichuki and by Fairhead and Leach. Both were methodologically complex and collaborative. Tiffen, Mortimore, and Gichuki examined landscape change in the Kamba region of highland Kenya.[6] In the 1930s, Ukambani had been a byword among British colonial administrators for environmental deterioration. Soil erosion was at an advanced stage, and vegetative cover was disappearing. Ukambani in the 1970s was a very different place. Soil erosion had been arrested and trees now proliferated. The local economy was vibrant and expanding. There were "more people" and "less erosion." Ukambani is not far from Nairobi. While their population grew exponentially, the Akamba people worked in the city and invested their wages in their farms back home to feed the city's market.

Fairhead and Leach had no less dramatic a story to tell.[7] Their point of departure was a long tradition of perceptions of deforestation in the former French colony of Guinée, ostensibly brought about by destructive, short-sighted local agricultural practices. Drawing on aerial photographs and remote sensing plus direct observation, informant interviews, and a documentary record, they concluded that the colonial and now postcolonial reading of the landscape had it backward. The landscape was properly viewed as a mosaic of savanna and forest. Far from the forest retreating, it was possibly advancing. Far from local practice being destructive of forest, it probably supported forest growth. At the heart of many forests they found villages, which they concluded were the seeds of the woods that grew up around them.

In the 1990s, research on Ethiopian landscape change reflected these strategies and motifs. James McCann reopened the discussion with the publication of his influential revision of Ethiopian landscape history.[8] He took up the question of deforestation, which he addressed with two specific examples: the open countryside around Ankobar in northern Shawa, a regional capital in the early and mid-nineteenth century; and Gera, in the upper Gibe River basin of southwest Ethiopia. The landscapes of northern Shawa and central Wallo are more or less continuous. McCann argues that the Ankobar region, in very broad terms, looks very much like it did in the 1840s, when it was visited by a number of British and French observers. There were no major forests and hence no deforestation. By contrast, Gera, where maize and coffee have been the principal agricultural products, has a more complicated history. Mid-nineteenth-century observers describe a patchwork landscape, densely settled in some places and elsewhere marked by forests, bush, and open fields. By the end of the century, the balance had shifted decisively toward the forest, which was taking over earlier farm fields. In more recent times, balance is again shifting back toward greater openness. Change has not been unidirectional.

In 1995 Halvor Wøien published an innovative paper on landscape change on the Mafud escarpment in northern Shawa, around the town of Dabra Sina. His work was based on aerial photographs that confirmed existing understandings of population increase and established a notable expansion of natural and woody vegetation.[9] His principal interest was in human impact. The agro-ecological zone is primarily *dagga*, and the farming practices are essentially the same as in Tahuladare. Wøien observed that the photographs showed an increase in the number of household compounds (signs of population growth) between 1957 and 1986, which varied according to agro-ecological zone, from 11.2 percent at higher altitudes to 45.6 percent in areas below 2,500 meters above sea level. Natural vegetative cover rose over the same period from 4.4 percent of the total land surface to 9.2 percent, and eucalyptus woodlots expanded from 1.1 percent of the total area in 1957 to 4.6 percent in 1986. Overall natural and woody vegetation increased from 5.5 percent of surface area to 13.8 percent. Much of this change Wøein attributes to conservation activities promoted by the Department of Agriculture—enclosure of hillsides, creation of eucalyptus plantations—but his maps may indicate private tree planting as well.

In 1998 the Ethiopian Wildlife and Natural History Society published a survey of the flora and fauna of the Simien National Park as a follow-up to Stähli's study. To document long-term landscape change, it used intensive field study along with matched photographic pairs. An important theme was human impact on wild flora and fauna, which it found to be negative—grazing cattle and other domestic animals had had "an absolutely devastating effect" on afro-alpine grassland (10 and figure 2.2 on 9); forests had been cleared (figures 2.5–2.18); other vegetation had been impoverished (figure 2.10); and the habitat of most larger mammals had been reduced. Mitigating these effects to a modest degree, farming activities had created higher diversity of habitat for small mammals (71 and 76), and while hardly in overall equilibrium, forests had expanded in some areas while contracting in others (figures 2.17–2.21). The conclusion left little space for local cultivators: "the overall human impact and especially the use of the grassland as pasture . . . have to be drastically reduced" (31). Its leading recommendation was that "the extensive impact of man and his livestock in the afroalpine ecosystems of the Gich plateau . . . must drastically be reduced"

(92). Informing the study was a patronizing attitude toward local farmers—"It is sad to see the warm and friendly Simen inhabitants are causing serious damage to their surroundings" (88).

In 1997 Alfons Ritler published the first of a series of studies that were to provide a meticulous account of observations made by European travelers between 1830 and 1930 of the landscape of north-central Ethiopia—Eritrea, Tigray, Bagemder, Gojjam, Wallo, and Shawa—the region of ox-drawn plow agriculture.[10] At the heart of his enterprise was a thorough reading of materials written in European languages.[11] In his PhD dissertation, he expanded his source material to include a few matched photographs. The observations made by travelers, although confined to the areas along historically established roads, can be located in space and time and provide a patchwork view of the landscape. Ritler's material indicates no traumatic changes in the landscape, which was visibly continuous with today's. The balance between cultivated and uncultivated did not fundamentally change, although the area under cultivation probably expanded. There had been no extensive forests in the mid-nineteenth century, and hence no deforestation, although almost certainly some reduction in the numbers of indigenous trees. He concluded that "landscape change is much more complex and long-term than has frequently been supposed."[12]

The turn of the century saw the appearance of a clutch of studies that further nuanced our understanding of change in Ethiopian landscapes. Most of them concluded that "deforestation" is an inappropriate category to describe landscape change, tree coverage has been a mosaic of gains in some areas and losses in others, and planting eucalyptus is of landscape significance. Gascon used personal observation with travelers' accounts and historic photographs to argue that tree cover varied greatly from region to region in Ethiopia and that the northern, plow-cultivating landscape had not been forested for a long time.[13] He explored the language of "forest," arguing that rather than distinct plant communities or formations, the Amharic terms—*chaqa* and *dan*—distinguished primarily between the intensively inhabited landscape and landscapes peripheral to them, that what is so frequently translated as "forest" might with equal justice be rendered as "bush," or "waste land."[14]

Boerma investigated woody vegetation on the Eritrean plateau, physically continuous with the central Ethiopian plateau and subject to the same cultivation practices.[15] She noted the divergence in Eritrea between perceptions of large-scale deforestation and the realities of change in woody vegetation. On the basis of archival research, informant interviews, onsite observation, and matched photographs, she concluded that change in tree cover was by no means uniform or unilinear but a kaleidoscope with both loss and gain in tree cover being experienced at different points in Eritrea's recent history (1). She explained the divergence between perceptions and reality with reference to the usefulness of the myth of large-scale deforestation to the various governments that controlled Eritrea from its creation in 1890 to the present. Deforestation has evolved into a powerful metaphor of past traumas and injustices suffered by Eritrea to the extent of constituting a collective memory to which everyone subscribes regardless of their actual experience of changes in tree cover (2).

Scheidegger's work presented case studies of ten highland landscapes ranging from Maychew in Tigray in the north as far south as Addis Ababa. The cases were based on eighteen matched pairs of ground photographs, the originals of which date mostly from the 1930s and the new photographs from 1998.[16] He found evidence of rapid population increase, but he also found "an increase in woody resources" (xiii), which he attributed to eucalyptus planting.

A collaborative project led by Reid and Kruska produced the most sophisticated and complex study of changes in farming and woody vegetation carried out in Ethiopia to date.[17] It focused on the Gibe River valley to the southwest of Addis Ababa and integrated data from aerial photographs (1957 and 1973) with data from satellite images (1987 and 1993) and ground observation and informant interviews. The multiple time points provided the researchers with examples of nonlinear change, which varied. The project used four land-use categories: smallholder cultivation, large-holder cultivation, riparian woodlands, and wooded grasslands. These categories ebbed and flowed across the landscape driven by such factors as land reform following the edicts of 1975; the advent of trypanosomiasis, which had a devastating impact on plow oxen and hence on the area under cultivation; villagization; and national drought, which caused immigration of farmers from other parts of Ethiopia.[18] The natural growth of the local human population was not an evident factor. This study underscored the complex and nonlinear character of landscape change in Ethiopia and the range of factors producing it. Changes in land cover and land use in eastern Wallo were as varied and complex as those found in other parts of Ethiopia.

Landscape Change in Eastern Wallo

Changing political conditions in Wallo affected farmers' lives and their farming decisions. Through much of the nineteenth century, politics in Wallo consisted of military struggles between regional lords for autonomy and expanded domination. In the earlier nineteenth century, these were local conflicts.[19] The national dominance of a dynasty from Yajjuto in the north of Wallo helped minimize the scale of conflict. Tewodros II (1855–68), and his eventual successor, Yohannes IV (1871–89), made the subjugation of Wallo central to their attempts to reconstruct what they understood to have been the national domain. Warfare in the province escalated and farmer insecurity rose. The advent of Menilek II in 1889 was followed by the Kefu Qan and then some fifteen years of internal peace. Wallo was governed by Menilek's protégé, Ras Mika'él, whose son, Iyasu, rose to the throne in 1913.[20] Three years later, he was deposed, plunging Wallo back into conflict with the powers in Addis Ababa, the new capital since 1888. This conflict is firmly established in the Wallo collective memory. Menilek's only surviving child, his daughter, Zewditu, succeeded Iyasu, her authority immediately compromised by the simultaneous appointment of Ras Tafari Makonnen, the future Haile Selassie, as "heir apparent" and, he successfully claimed, "regent." Mika'él rebelled in support of his son and was defeated in the Battle of Sagale in November 1916. Meanwhile, Iyasu remained at large and mustered an army against Dajjazmach Habta Giyorgis, who had occupied Wallo on behalf of the national government following Sagale. On August 27, 1917, this army was again defeated, following which Habta Giyorgis's men "viciously pillaged" Wallo.[21]

Up to 1935 Wallo was ruled by a succession of unpopular appointees from Shawa. In 1935 the Italians invaded and decisively defeated the main Ethiopian army at the Battle of Maychew in March 1936. Dessie was an important staging point for the northern front. The emperor took the road north through Dessie, and he retreated back through Dessie to Addis Ababa following defeat at Maychew. Dessie became an important regional center for the new Italian colonial administration. The Italian occupation left few marks beyond the layout of towns like Dessie. On his restoration in 1941, Haile Selassie appointed a new governor of Wallo: Asfaw Wossen, his eldest son, whose mother was Empress Menen Asfaw, daughter of a longstanding dynasty of the Wallo district of Ambassel. Asfaw Wossen's rule has been seen as indolent. The three decades of the postwar era, generally viewed as comparatively prosperous, came to an abrupt end in the drought and famine of 1972–73. Wallo was hit particularly hard, its farmers blamed for landscape practices, which, if they did not cause drought and famine, certainly contributed to their onset. What landscape evidence is there to support this view?

A starting point for understanding landscape change in Wallo was set by a scattered array of European travelers and a British military expedition in 1868. Seventeen European observers passed through this general area in the century between 1830 and 1930. Their insights remain of value. In later years, they increasingly passed through Tahuladare along the route past Lake Hayq and Dessie. The landscape was much less densely populated than it has since become. Ritler's summary of their observations is that there was extensive heavy brush and many trees. (See table 5.1 for a listing of Ritler's sources for the period 1830–1930.) Most woodland was found along the rivers, on hills, and on mountain ridges.[22] Particularly notable was heavy forest in the Mille Valley just south of Wechale. No forest was observed in the *dagga* agro-ecological zone, but there were reports of "extended wooded areas." The area around Lake Hayq and on the hills above it was one of a number of areas with "an abundance of trees" (38). But equally there were large tracts of woodless country. One traveler noticed the particularly heavy forest, which still lies southeast of Dessie. To illustrate his account of the years 1865–1930, Ritler reproduced a photograph from about 1910 of the palace of Ras Mika'él at Dessie and its surroundings.[23] The vista is of a treeless meadowland, which, landforms and the building aside, is unrecognizable in today's landscape, which is heavily wooded with eucalyptus trees.

Ritler and Scheidegger drew on photographs made by the British expeditionary force, which invaded Ethiopia in 1868. This force marched from the coast through Tigray to Maqdala, the fortress and administrative center established by Emperor Tewodros, where he had imprisoned a number of foreigners, missionaries under British protection prominent among them. Maqdala is in Wallo, perhaps seventy kilometers northwest of Dessie. A photographic team was part of this expedition. Its main job was to produce multiple copies of documents, such as maps drawn up for the following day's march from information gathered by advance scouts. The photo team would spend much of the night producing copies of these maps for the use of the troops. Only incidentally did they take the earliest photographs of people and landscapes in Ethiopia. These photos were later made available in albums, of which quite a number have survived.[24] Apparently

Table 5.1. Travelers' accounts of southern Wallo, 1830–1930

Period covered	*Name*	*Title*
1835–36	E. Combes and M. Tamisier	*Voyage en Abyssinie, dans le pays des Gall de Choa et de l'Ifat* . . . , 4 vols. (Paris, 1838)
		1840s
1842	J. L. Krapf	C. W. Isenberg and J. L. Krapf, *The Journals of the Rev. Mssrs. Isenberg and Krapf . . . Detailing their Proceedings in the Kingdom of Shoa* . . . (London, 1843)
1843	T. Lefebvre	*Voyage en Abyssinie execute pendant les annés 1839* . . . (Paris, 1845)
1843	Arnaud d'Abbadie	*Douze ans de séjour dans l'Haute-Éthiopie*, 3 vols. (Vatican, 1980), volume 1 first published Paris, 1868
1844	W. C. Plowden	*Travels in Abyssinia and the Galla Country* . . . (London, 1868)
		1860s
1860	H. A. Stern	*Wanderings among the Falashas in Abyssinia* . . . (London, 1862)
1862	T. von Heuglin	*Reisen in Nord-Ost Afrika, Tagebuch einer Reise von Chartum nach Abessinien* (Jena, 1868)
1866–68	H. Rassam	*Narrative of the British Mission to Theodore, King of Abyssinia* . . . , 2 vols. (London, 1869)
1868	G. Rohlfs	*Im Auftrage Sr. Maj. des Königs von Preussen mit dem englischen Expeditions-Corps in Abessinien* (Bremen, 1869)
		1870s
1877–81	A. Cecchi	*Da Zeila alle frontiere del Caffa* (Rome, 1886–87)
		1880s
1881	J. V. Abargues de Sostén	*Noticias acerca de la expedición cientifica, geográfica y mercantile realizada en el África Oriental* (Madrid, 1883)
1882–84	P. Soleillet	*Voyage en Abyssinie* (Rouen, 1886)
1896	A. B. Wylde	*Modern Abyssinia* (London, 1901)

Table 5.1. (*continued*)

		1900s
1901–11 (?)	L. de Castro	*Nella terra del Negus*, 2 vols. (Milan, 1914–15)
1908–9	J. Faitlovich	*Quer durch Abessinien: Meine Zweite Reise zu den Falaschas* (Berlin, 1910)
1901–10 (?)	A. Cipolla	*Pagine africane di un esploratore* (Milan, 1927)
1910 (?)	C. A. Annaratone	*In Abissinia* (Rome, 1914)

Source: Ritler, *Land Use, Forests and the Landscape of Ethiopia*, 38–39, 185–96; Ritler, *Forests, Land Use and Landscape in the Central and Northern Ethiopian Highlands*, 88–92; Ritler, *Anhang zu Wald, Landnutzung und Landschaft im Zentralen und Nördlichen äthiopien*, 256–67.
Note: Scheidegger reports his findings according to current administrative organization, by which south Wallo is the "zone," which encompasses "eastern Wallo" as used below, and, within it, Tahuladare.

the negatives have not survived. The photos used by Ritler and Scheidegger were different views of the landscape around Lake Ashangi in southern Tigray directly north of Wallo.[25] Having analyzed the historic photographs of Ashangi, a number of them later than 1868, Scheidegger notes that from our earliest record the land was already under intensive agricultural and grazing use, trees were scarce, and the pattern of land use is continuous to the present. He also noted a more intensive use of the land and population growth evident in an increase in local settlements.

To date the most ambitious use of the Maqdala photographs has been by Jan Nyssen and colleagues, who rephotographed and systematically analyzed thirteen of the pairs.[26] Table 5.2 shows the indicators of landscape change and the ratings based on a five-point scale. The photographs document landscapes in southern Tigray and northern Wallo, areas physically continuous with my study area and under comparable patterns of land use. Nyssen and colleagues deemed that continuity, rather than trauma, characterized landscape change. Eucalyptus trees, unknown in Ethiopia in 1868, had made an impact on the landscape by 2008. Nyssen and colleagues assessed change in vegetation by separating out the eucalyptus from the indigenous trees. They found that planting eucalyptus improved overall landscape health and had no negative effect on the state of indigenous vegetation, which seemed to have maintained itself. This finding, which contradicts some expert views, is plausible and in keeping with my own findings.

TABLE 5.2. Interpretation of Ethiopian landscape change between 1868 and 2008 as evidenced by rephotographing selected photos of the Maqdala expedition

Indicators of visible soil erosion	*#*	*Score*	*Change*
Woody vegetation	13	1.2*	Improved
Woody vegetation except eucalyptus	13	0.3^{ns}	Slightly improved
Visible soil erosion	12	0.1^{ns}	
Soil and water conservation	12	0.7*	Improved
Land management	13	1.0*	Improved

Source: Adapted from Nyssen et al., "Desertification?," 2752, table 1.
Notes: # = number of landscape sites. Score = average of median scores given to all interpreted landscapes from -2 (strong deterioration) to +2 (strong improvement) with level of significance for the deviation from a test value zero (= no change) (*significant at 0.05 level; ^{ns} not significant). Change = 2008 compared to 1868.

1937: A Baseline for Twentieth-Century Wallo Landscape Change

There are several good accounts of the Italian occupation of Ethiopia, but little information about their activities in Dessie and its environs.[27] They established Dessie as the capital of "Uollo," one of the six *commissariato* within the "Governo" of "Amara," the capital of which was at Gondar. They left a lasting mark on the town in creating its central piazza. Within the larger framework of a policy to encourage economic development through farms run by corporations and European settlers, they granted a number of agricultural concessions in the Dessie area—in the Garado Valley, to the south of the city, and at Boru Meda and Alasha Meda, to its northwest—which have left no visible trace on the landscape and are remembered today only by the elderly.

The Italian colonial administration carried out an inventory of the region's "scarce" woody resources. The report estimated that woodland covered some 800,000 hectares or 20 percent of the territory of Wallo, much of this bushland and only part of it forested in a strict sense. Within Wallo were only three "forests" worthy of note: Danqoro in Borana and Yagof and Albuko southeast and south of Dessie.[28] The Danqoro and Yagof forests continue to exist, but only vestiges of Albuko remain. The Italian report estimated Danqoro to cover about 700 hectares, Albuko only 150, and gave no estimate for Yagof.[29] While they imagined Dessie would become one of the most dynamic economic centers of Italian East Africa, its woody resources were insufficient to meet the need for construction wood, although it might suffice for tanning and firewood. The greatest risks to existing woodland came from fire and "premature and irrational cutting," from using woodland to pasture livestock and from shifting cultivation.[30] The Italians also fancied that the "atmosphere of tranquility and security," which they had introduced and was "the first the area had ever experienced," might encourage an expansion of land under cultivation at the expense of bush and forest. This threat could be warded off through regulation, propaganda directed to the "chiefs," and the threat of force from a Native Forest Corps (Corpo Forestale Indigeno), along with the establishment of "settled agriculture."

The Italians published very little about Wallo during their occupation. The most enduring contribution was Brielli's account of historical traditions, which has influenced numerous subsequent accounts of Wallo history.[31] Alessandro Lusana published a substantial report of natural resources in eastern Wallo, based on a trip made in May 1937.[32] Lusana went north from Dessie to the highlands east of Hayq, then made a long journey southeast to the market town of Bati. From there he headed south, returning to the main Addis Ababa–Dessie highway ("la strada imperiale") some distance south of the BorQanna River. The early part of this route took him through or very near two of the peasant associations—Gwobeya and Gerba—which were later focuses of the MacArthur-funded collaborative research project. His trek took two weeks and would be challenging to duplicate today because only its last leg used an all-weather road.

Lusana provided a narrative account of the trip followed by detailed information on mountains, hydrography, climate, economic resources, and ethnography and sociology. A major goal was to assess the suitability of the region for settlement by Italian farmers. Judgments of the country that Lusana passed through were generally positive—"densely populated," "rich," "pretty," "many plowed fields." He hoped his report would do justice to "a region, which, because of its possibilities and fortunate location is destined to become one of the most noteworthy of the Empire."[33] His summary account of "arboreal vegetation" is that the region was "without woods," "stripped of trees" as a result of its dense population.[34] He also noted that "the importing of Eucaliptus, judging from the exuberance and rapid growth of existing woodlots in Dessie, has brought optimal results."[35]

The Italian occupation gave rise to one record of the landscape of eastern Wallo of enduring value: a set of photographs taken by Armando Maugini, director of the Istituto Agronomico dell'Africa Italiana in Florence.[36]

Maugini seems to have been first and foremost a technocrat who became deeply involved in shaping Italian land and agricultural policy in the African empire, first in Libya and then in Ethiopia after 1936.[37] He visited Dessie at least twice, in February 1937 and February 1940. He was an enthusiastic photographer whose cameras included a Leica and a Voigtländer.[38] In 1940 he visited Italian concessions in the Garado Valley and at Boru Meda and Alasha Meda. In 1937 he traveled more widely. On February 16 he landed at the airfield in Kombolcha, which still serves Dessie. He drove eastward past Kombolcha and along the road, which had recently been built and still runs from the town to Bati and Asab.[39] He photographed the airfield, Kombolcha town, landscapes along his route, and Bati and its market. He passed close to the Gerba peasant association, after which he returned to Dessie, where he spent the night. Part of the next morning he devoted to photographing the town and then set out on the main highway north toward Asmara, stopping along the way to take photographs. This took him through the heart of what is now Tahuladare. His itinerary covered more ground than this and the complete set of photographs from his multiple trips cover Eritrea, Tigray, Gondar, Shawa, and Harar, in addition to Wallo.[40] We have to infer the rationale for the 1937 trip, since I could not find a diary, log, or report documenting it. Given Maugini's formal position, his involvement in agricultural development policy, and the subject of most of the photographs, it can safely be assumed that he was canvassing agricultural potential, possibly for the settlement of Italian farmers. Three years later he returned to visit the concessions around Dessie.

The Maugini Photographs

The Maugini collection contains hundreds of prints in chronological order, dated and identified by distance from the main point of departure for the day, in this case Dessie and Kombolcha. The negatives appear not to have survived, so the photos are available to us as prints, some as enlargements, others as contact prints. Although their quality is usable, obviously the enlargements contain more information than do the contact prints, which are much smaller in size. The detail in some of the contacts is limited. I copied the photos twice, in June and again in December 1994. Altogether I copied ninety-one photographs of eastern Wallo, the originals of which had been taken on February 16 and 17, 1937.[41] They document the terrain along the highway from Kombolcha to Bati and from Dessie to

Wechele, as well as the town of Dessie. Of these, thirty-eight proved to be replicable and of value in recording landscapes relevant to the collaborative research project that focused on the three peasant associations in Tahuladare. None of the photos was taken within one of the project's peasant associations, but some record immediately adjacent landscapes. The photographs became the point of departure for interviews I conducted with elder informants, who were old enough to have lived through the landscape changes evident in the paired photo. These interviews are the basis for chapter 6.

The Maugini photos give us as representative a view of the landscape of eastern Wallo in 1937 as we are likely ever to get. They have obvious limits. All of them exclude far more than they include; they are driven by the interests of the person who took them; and they were taken from a highway in a place where roads are still scarce. These limits have compensating merits: exclusion focuses attention; Maugini's interests are relevant to landscape history; and the road is likely to have had only modest landscape impact, since it had been completed only a year or two before the photographs were taken. Although that impact undoubtedly intensified over the following decades, this in no way compromises the value of the 1937 photos, and we will see that road effects may have been complex. Awareness of their limitations ought not to impede reading the photographs for the evidence that they do contain.

To get a sense of the Wallo landscape as Maugini saw it, we would do well to look at a few of the photos in the order in which he took them—first Kombolcha and the road down to Gerba, then Dessie, and finally the road north from Dessie. Figures 5.1 and 5.2 portray the landscape of the Kombolcha plain, looking down from the Muslim cemetery on a hill east of the town. They reflect Maugini's common strategy, which was to take two photos from the same spot by pivoting roughly 90 degrees. These photos have been cropped from my negatives to include only the Maugini original, which was an enlargement. The foreground of both photos shows euphorbias, protected by being in a cemetery. Figure 5.2 shows a well-vegetated hillside in the right middle distance. Both photos suggest that the Kombolcha plain in 1937 was virtually devoid of trees, the darkness on the valley floor in figure 5.2 being cloud shadow.

Figure 5.3 shows a scene along the Dalaqa River, ostensibly twenty-two kilometers from Kombolcha on the road to Bati, but by my reckoning, more like ten kilometers. The original is a contact copy. In the middle of

Figure 5.1. "La pianura di Combolcià"
Source: Maugini (1937)

Figure 5.2. "La pianura di Combolcià"
Source: Maugini (1937)

Figure 5.3. "a 22km da Combolcià"
Source: Maugini (1937)

the photo is an Italian road works. While because of cloud shadow, parts of the photo are too dark to allow definitive judgment, the landscape appears patchy with shrubland alternating with open land, some of it farmed. The steepness of the slopes on the right apparently set a limit to tilling, although just a little above the center of the photograph there appears to be a cultivated stretch running up a rather steep slope. There would appear to be terracing on this stretch. Few trees are clearly identifiable.

Figures 5.4 and 5.5 show the town of Dessie looking southwest from Aytayaf, the palace of Ras Mika'él, onto the lower town and the road to the Garado Valley and then more westerly. Figure 5.4 shows a bare mountain on the right and then a fringe of eucalyptus at its base toward Garado on the left and in the foreground at the base of the hill from which the photograph was taken. Figure 5.5 shows an equally bare mountain in the background, several stands of eucalyptus in the right at the foot of the mountain, and more eucalyptus on the valley floor below the camera's position as well as more directly below in the heavily shaded foreground. Several of the eucalyptus trees are mature and must have been planted well before 1935.

Figures 5.6 and 5.7 are another example of Maugini's strategy of taking at least two photographs from the same position by rotating 90 degrees.

Figure 5.4. Entrance to the Garado Valley
Source: Maugini (1937)

Figure 5.5. Dessie from Aytayaf
Source: Maugini (1937)

Figure 5.6. "Vallata a 6 km oltre Dessiè," looking toward Kombolcha from the north
Source: Maugini (1937)

Figure 5.7. "Vallata a 6 km oltre Dessiè," just north of Bar Qorke village toward Kombolcha
Source: Maugini (1937)

Figure 5.6 looks eastward and 5.7 looks roughly north. These farming landscapes were the first Maugini roadside sites I was able to identify. The landforms remain quite recognizable. Figure 5.7 was probably taken fairly early in the morning, hence the puff of smoke hanging over the village. The village and its immediate vicinity are fairly dense with woody vegetation, while the rest of the landscape is generally bare and marked out in agricultural fields, although there is bush on the hill opposite the village on the right and above it on the left. Figure 5.6 shows a deep valley, its floor and lower slopes marked out in farm fields, its middle and upper slopes covered with brush. The only tree, which can be unequivocally identified as such in either illustration, is on the middle horizon of figure 5.7.

Figure 5.8 provides a broad view of a hillside across the Mille Valley from a spot on the road some distance beyond the town of Passomille. The Mille can be seen in the lower left corner. Although the reproduction is not optimal, it is still possible to identify the main features. This is a cultivated landscape, well settled. Apart from the bare mountain slopes on the far horizon, the landscape is a mix of cultivated (generally the lighter areas) and brush. There is a patch of brush in the lower foreground; again to its right and left; two stretches of bush in the middle of the photo's left side;

Figure 5.8. Mille Valley
Source: Maugini (1937)

and finally, a stretch on the right of the photo at the foot of the mountain behind. As for trees, there might be three or so on the right of the photo below the brush at the foot of the mountain.

Finally, figures 5.9 and 5.10 are more photos from the same position and they form a panorama with the mountains of Ambassel to the south in the background. The foreground is part of the broad plain of the Mille a little to the north of the town of Chis Aba Lima. There is no sign of farms or houses. The land might well have been used for grazing, but there are no cattle, sheep, or goats in view. The vegetation is uncultivated and rather sparse with clumps of grass, bushes (both isolated and in small clumps), and three or four trees. It is the landscape emptiest of obvious human presence, which Maugini photographed between Gerba and the town of Wechale, which lies just to the north above the plain. By Maugini's reckoning we are now about forty-five kilometers north of Dessie on the road to Asmara, although we might be closer to fifty kilometers away.

From the mid-nineteenth century to 1937, the Wallo landscape was a settled, human one. Its population was somewhat scattered, especially in the nineteenth century when political insecurity encouraged strategic concentration. Farming and herding were the principal occupations. Europeans viewed the farmed landscape with some degree of favor, although contradictorily sometimes disapproving of Ethiopian farming practices. Erosion, occasionally mentioned, did not force itself on observers. Change over this period is impossible to assess. We do know that rinderpest must have brought desolation and a contraction of the human imprint in the 1890s and, probably, into the early twentieth century. Other livestock diseases would have had their impact, although none so dramatic as the rinderpest. We have no information specific to Wallo concerning human disease, although we have seen that the Great Famine elsewhere in the Ethiopian highlands fostered hostile microbes and sapped human mortality. So it must have been in Wallo. We can be certain that climate variation from season to season and year to year provided a pulse for the expansion and contraction of area under crop and pasture and shaped decisions about what to plant and where. This was not a forested landscape, although there were a few forests and concentrations of trees along the water courses. Rephotographing the Maugini landscapes in 1997 suggested continuity with the past; showed little evidence of trauma, although we know such to have happened in the famines of 1973 and 1984; and illustrated the growth of human population as established by the census of 1984.

Figure 5.9. "A 45 km da Dessiè," Mille plain north of Chis Aba Lima
Source: Maugini (1937)

Figure 5.10. "A 45 km da Dessiè," Mille plain north of Chis Aba Lima
Source: Maugini (1937)

Maugini Rephotographed

In February 1997 I rephotographed the Maugini photographs sixty years to the month after they were originally taken. The following selection of paired photos documents the landscape along a corridor running from the Dalaqa Valley east of Kombolcha to the plain of the Mille River, some fifty kilometers north of Dessie. The total distance represented is approximately 100 kilometers. I chose the pairs partly according to the legibility of the original and the richness of the comparison of the original to the rephotograph and to illustrate distinctive landscapes. The selection does not give proportional coverage to the area's three main agro-ecological zones—*wayna dagga* is overrepresented, *dagga* underrepresented, and *qolla* (lowland) is absent—although it does reflect the proportional coverage of the original corpus. I take them up in the order in which Maugini photographed them, the original side by side with the rephotograph with an eye to changes in the agricultural landscape. Knowing that they had all been taken from the highway, together with the multiple dimensions created by Ethiopia's mountainous landforms, made locating the originals fairly easy, once I had identified the first of the originals. Exact reproduction proved difficult, although many of the new ones are close to the originals. In a few cases realignment of the road made reproduction impossible—and these failures are not reproduced here. In other cases an increase in foreground foliage made close approximation difficult, even when I was confident I had found the spot from which Maugini had originally taken the photograph. As a body the photo pairs provide a sixty-year framework for approaching the twentieth-century history of the eastern Wallo landscape.

The Kombolcha Plain

We start with a broad perspective on the plain on which the town of Kombolcha sits. The view in figure 5.11 is to the west or southwest. On the left is the base of Yagof Mountain and its protected forest. Dessie lies out of sight in the hills forming the background. In the immediate foreground are euphorbias, flourishing in the protected environment of a Muslim cemetery. In 1937 the valley floor was devoid of noticeable vegetation, apart from grass and the possible exception of brush along a river course, which cuts slightly down to the right in the middle of the photograph. Were it not for the common horizon in the photographs, it would be difficult to recognize the 1997 photograph as having been taken from the same vantage point.

Figure 5.11. "La pianura di Combolcià"
Source: Maugini (1937, *left*); Crummey (1997, *right*)

Although there appears to have been some loss of euphorbias in the foreground, the valley floor now shows extensive eucalyptus planting. The town of Kombolcha has clearly expanded.

Figure 5.12 shows the second view of the Kombolcha plain in 1937 taken from the same position as the previous photograph, but on a pivot to the right, toward the north, showing a plain largely devoid of woody vegetation, although the distant hills, most clearly the nearest one, show brushy vegetation. The immediate foreground is sparsely populated by grass and euphorbias. The hillside to the right is well vegetated with euphorbias and bushes. I chose the result of my first attempt in May 1996, to rephotograph 7169/190, rather than the 1997 version. Although the shift of three months would have had some effect on the vegetation, the 1996 version is clearer than the 1997 version. In the foreground the euphorbias are richer than they had been in 1937, but the front of the hillside to the right has been

Figure 5.12. "La pianura di Combolcià"
Source: Maugini (1937, *left*); Crummey (1996, *right*)

stripped, while the upslope vegetation has fared pretty well. An extension of the eucalyptus woods in a 1997 photo is evident on the valley floor and in smaller stands on the more distant reaches of the plain. Neither attempt at reproduction contains any detail on the mountains in the background.

The Dalaqa Valley

Figure 5.13 shows us the narrow rugged valley of the Dalaqa River downstream from Kombolcha. In the middle of the photograph is a roadworks site, of which no trace remains. Scattered fields indicate that people are farming here, although they are constrained by the steepness of the terrain. The 1997 photograph was taken from a little higher up and a little further back and shows more of the hillside to the right of and immediately above the river in the right middle foreground. Its furthest reach shows devegetation and erosion, although the nearer reach remains well vegetated. Terraced cultivation now reaches further up the side of the hill forming the middle horizon, and its sides are less well vegetated. A church sits atop the knoll in the upper left quadrant. Vegetation around it seems little altered over the sixty-year interval, whereas just before it more land appears under cultivation.

The View from North of Dessie

Driving north from Dessie on the morning of February 17, 1937, Maugini passed a village through a narrow gap, beyond which the landscape suddenly opened. The road turned sharply to the left to skirt a precipice overlooking a valley, which fell away eastward to the right down to the Kombolcha plain; at the head of the valley, still several hundred feet below

Figure 5.13. "A 22 km da Combolcià Cantiere Puricelli"
Source: Maugini (1937, *left*); Crummey (1997, *right*)

and a little to the left, sat the village of Qobbo in what is now the Qorke peasant association.

Figure 5.14 looks along the valley descending to the Kombolcha plain, part of the headwater drainage of the BorQanna River, a tributary of the Awash that flows through Kombolcha. The valley floor was intensively cultivated, as were the lower slopes of the valley walls. Difficult to make out from the reproduction of the photograph is the state of the steep slope in the right foreground. In 1937 it was healthy and well vegetated. In 1997 its vegetation was reduced and soil erosion was under way. To the left, the hillsides of the valley in 1937, in the uppermost part of the visible valley, were cultivated, while further east the vegetation would seem parallel to that in the immediate right foreground. By 1997 much of the left foreground was no longer in cultivation. Woody vegetation has replaced it, probably through a government-sponsored afforestation scheme. Barely discernible is continuity in the shape and location of individual farm fields. Such continuity is clearer in photographs we consider later.

Meanwhile, Qobbo village, to Maugini's left—figure 5.15—sat amid farm fields in a landscape largely devoid of woody vegetation, although there was some bush above the village to the left of the photograph and in the photo's left foreground. A solitary tree sat on the horizon of the hill behind and above the village. The village itself was so densely vegetated in 1937 that it is impossible to count the number of dwellings it contains. Whatever their number, by 1997 they had multiplied and the village had more than doubled in size. Numerous eucalyptus trees are evident within the village, along the watercourse that runs down into the photo's lower right corner, on the hills on the right of the photograph, and less densely on the hills

Figure 5.14. "Vallata a 6 km oltre Dessiè"
Source: Maugini (1937, *left*); Crummey (1997, *right*)

that mark the middle horizon. It is very difficult just from the 1997 photograph to discern any vegetation in the village other than eucalyptus trees and some bushes, but a visit to the village revealed more varieties. I spoke to six elders—four men and two women. All were planters of eucalyptus. Four had planted an additional kind of tree—juniper, *besana* (*Croton macrostachys*), or the euphorbia *qulqwal*. One resident had planted five different kinds of trees in addition to the eucalyptus—peach, orange, juniper, *besana*, and *sombo* (*Ekebergia capensis*).[42] The big stand of eucalyptus on the hillside on the right was probably the result of a government-planting scheme. Less readily seen but established by local interviews was a privately owned eucalyptus plantation that juts down a little from the rest of the trees. In the immediate foreground the hill looked very stressed, denuded of vegetation.

Figure 5.15. "Vallata a 6 km oltre Dessiè"
Source: Maugini (1937, *left*); Crummey (1997, *right*)

Sulula and the Dadaho River Valley

A few kilometers north of Qobbo village, thirteen kilometers from Dessie by his reckoning, Maugini entered the valley of the Dadaho, a little south of the village of Sulula. It attracted him. He took seven photographs within a space of five kilometers, the first two—figures 5.16 and 5.17—just to the south of Sulula, which appears in the rear left corner of the rephotograph. The next five he took a little to the north of the village. Rephotography of some of these proved challenging because of increased foreground vegetation. He labeled the first two photographs as "Italcià," which refers more properly to the valley north of Sulula.

Maugini's first view northward along the valley revealed an open landscape bounded to the east by steep hills. Revisiting the area in 1997 allowed

Figure 5.16. "A 13 km da Dessiè. Italcià"
Source: Maugini (1937, *left*); Crummey (1997, *right*)

Figure 5.17. "A 13 km da Dessiè. Italcià"
Source: Maugini (1937, *left*); Crummey (1997, *right*)

a fuller appreciation of the original photograph, in which a lonely tree sits on the far horizon. The valley floor in the foreground is meadowland, subject to periodic flooding from the river, which runs along the right at the foot of the hill. Further back along the valley floor may be cultivation, which is almost certainly the case with the lower slopes of the hills. In 1997 the foreground vegetation had changed little, but the horizon on the left and the hillsides to the right both show eucalyptus plantations. The one on the left may be part of a government tree nursery, but the one on the right is a private woodlot.

The second photo of this set allows a fuller understanding of the landscape, since it was taken at a right angle to the previous photograph, now looking eastward at the hillside. In 1937 the fields at the bottom of the hill were tilled and there were tilled fields further up, around the settlement. Toward the top of the hill, two fields were visible, and there may have been two more smaller fields below them to the left. Apart from the fields and the houses, the landscape is marked by bush. By 1997 all cultivation had been eliminated from the hill above the settlement. It was covered with densely planted eucalyptus, part of a government afforestation project. The area under cultivation on the level of and below the settlement has expanded. Most of the brush on the hillside below the settlement has been eliminated in favor of terraced cultivation.

Just north of Sulula, Maugini took a series of photos that showed the descending ridge on the east side of the Dadaho Valley. The first, figure 5.18, which he reckoned to be fifteen kilometers from Dessie, had farm fields in the foreground, a threshing floor midway back to the right of the photo's center, and receding ridges in the background. Much of the plain was taken up by farm fields. There was a watercourse, with scattered trees and bushes, running from left to right across the middle of the photograph. On the right, perhaps three-fifths of the way up the photo, was a hillside with woody vegetation. At its bottom runs the Dadaho, the course of which was evident again in the line of woody vegetation, which ran just at the base of the background ridge. The side of the ridge facing me appeared to be mostly given over to cultivation, with scattered bushes. The top of the ridge again had woody vegetation, some of it probably surrounding farm settlements.

Sixty years later, the farm field in the foreground was easily recognizable. The rows of stones marking out separate plots appear not to have changed. Much of the woody vegetation along the watercourse running from left to right had been replaced with eucalyptus, although a dense growth of acacia had developed just to the right of the center of the photograph. The northernmost reach of the ridge was no longer visible, because of the tall stand

Figure 5.18. "A 15 km da Dessiè"
Source: Maugini (1937, *left*); Crummey (1997, *right*)

of eucalyptus just to the left of the acacia grove. The largest hillside was still visible. There was more vegetation along the crest of the ridge and bush had expanded just below the ridge. Below that it looked as if the area under cultivation had expanded, although it is difficult to judge to what extent. The hillside fields were terraced. The banks of the Dadaho, visible on the photo's right, a little over halfway up the side, were now sharply eroded, as were the banks of the watercourse that runs down from left to right, separating the near farm fields from those just below the line of eucalyptus. Difficult to see, but just below the erosion on the Dadaho was an iron telephone pole, present in 1997 as it had been in 1937.

A View across the Mille

Maugini had crossed the Mille, and by his reckoning had traveled forty-two kilometers from Dessie, when to the east there emerged a broad landscape across the river and running up the mountain side (figure 5.19). At the very bottom of his photograph he included the Mille. Farm fields covered much of the mountain, with meadow land and stretches of bush also visible. The banks of the Mille at the bottom of the photo were generally covered with brush, of which there was a large patch on the left side of the photo above the river. Immediately above that there was farming with another sizable patch of brush above it. Finally, on the right of the photo, at the foot of the final rise of the mountain, there was again a patch of brush. There may be a few trees on the right side of the photo at the head of the river gully and below the brush at the foot of the mountain.

In 1997 the pattern of field and bush was much as it had been in 1937. The fields retained the outlines, and there was still brush in most of the

Figure 5.19. Across the Mille
Source: Maugini (1937, *left*); Crummey (1997, *right*)

places where it had been in 1937. The 1997 photo also revealed change, not least at the bottom, where the Mille has apparently much broadened its bed at the expense of meadow land. There was a loss of brush in the lower right corner, but following the bank of the river to the left, we find a patch of brush much as it had been, decidedly thinner but possibly covering more area. Two fields perceptible within the 1937 photo seem to have largely disappeared. The large stretches of brush on the left of the 1937 photo covered much the same area, although again possibly thinned. Even the brush land at the top right of the 1937 photo was still evident in 1997, although there was now a sizable settlement where none was visible in 1937. It is quite possible that in 1997 one could still identify a solitary tree in a farm field on the edge of the photo about one-third of the way up.

The Mille Plain

A few kilometers beyond the road emerges from some low hills to run about five kilometers straight across a plain from which it then climbs to Wechale. On the far side of this plain lies the town of Chi Aba Lima, and just beyond the town, about forty-five kilometers north of Dessie by his calculation (perhaps more like fifty), Maugini hit on a striking formation in the Ambassel mountain range bounding the valley to the west. He took two photos, shown in figures 5.20 and 5.21, which fit together to create a panorama. In 1937 there was no sign of a human presence—no farm fields, no evident pasture, and certainly no animals or people. The basic vegetation would appear to have been grass, across which were scattered bushes and some trees, the bushes sometimes isolated, sometimes in clumps. By 1997 this landscape had undergone the most complete transformation of any of

Figure 5.20. "A 45 km da Dessiè"
Source: Maugini (1937, *left*); Crummey (1997, *right*)

Figure 5.21. "A 45 km da Dessiè"
Source: Maugini (1937, *left*); Crummey (1997, *right*)

those documented by Maugini anywhere in Wallo. It is the clearest evidence we have of expansion of the area under cultivation at the expense of other kinds of land cover. By 1997 almost the entire plain was under sorghum cultivation. The repeat photograph was taken following harvest, and the sorghum was not much in evidence, but just off camera were numerous stacks of sorghum stalks. Farmers can also be seen plowing the fields in preparation for sowing before the next agricultural season. All the brush and trees visible in 1937 have disappeared to reveal behind them a small number of trees scattered across the cultivated landscape. The farmers who work these fields live in settlements spread along the foot of the mountains. The history of the landscape is contested. I interviewed three elders from the settlements who averred that in their childhood in the 1930s the plain was already cultivated "all the way to the road." The photos suggest otherwise. Moreover, this is part of the Mille plain, part of which, in travelers' accounts and more recent testimony, was heavily forested, presumably closer to the river behind the photographer.[43] About 1,600 meters above sea level, the plain lies in the *dagga* agro-ecological zone. From the 1950s onward, farmers in northern Ethiopia, at least in northern Shawa and Wallo, were moving downslope into the *dagga* primarily in response (it would seem) to population pressure in the highlands.[44] The settlement of the Mille plain, farmer testimony notwithstanding, would illustrate this movement.

The Photo Pairs Summed Up

The photographs taken by Maugini in 1937 dovetail with the photographic and written evidence from the mid-nineteenth century onward and allow us to draw a more detailed, concrete, focused picture of the landscape of eastern Wallo. It is a vegetated landscape, the product of tropical wet and dry seasons. It is a human landscape subject to the needs of farmers and herders. Pasture was important to farmers and herders, the former dependent on the draft power of oxen and given to investment in small stock. From the photographs it is difficult to estimate the amount of land under the plow, but taking all of the photos together, it seems reasonable to suppose that this may have been around 50 percent. It is equally difficult to estimate how much of the plow land was planted in crops at any given time and, conversely, how much was under fallow, a practice we know to have been widespread if not universal in 1937.

When the Maugini originals are set beside photographs of the same landscapes sixty years later, numerous changes are evident. Population has grown. Although there may have been a corresponding expansion of land under cultivation, this would appear to have been modest, except for a large expanse of the Mille plain, uncultivated in 1937, which by 1997 had been entirely brought under sorghum cultivation. Missing from the photographs reproduced here are large stretches of meadow land on Boru and Alasha Meda, which Maugini documented. Elsewhere, dedicated meadow land is less easy to observe, although there is some in the foreground of figure 5.16 and, less obviously, scattered in figures 5.18 and 5.19. We may suppose that the hillsides provided opportunities for grazing and this would be one source of the stress evident in the hillsides most easily observed in 1997.

All sources down to 1937 suggest that woody vegetation was modest in quantity. The Italian review of forest resources stressed their limitations. None of Maugini's photographs registered large stands of indigenous trees. Indeed, some of his landscapes appear almost treeless. Woody vegetation in 1937 was composed mostly of brush, which by 1997 may have thinned. One of the most obvious changes between 1937 and 1997 is the proliferation of eucalyptus trees. Some of this was the product of large-scale government afforestation projects throughout the 1980s. Close inspection of the photographs suggests that farmers may also have been active in planting eucalyptus of their own accord. Government schemes were almost invariably large-scale, whereas smaller plantings, around homesteads and scattered across the landscape, were almost certainly the work of individuals. Less obvious are indications that farmers were also planting other woody plants. Forests being conspicuously absent at any point in the record as far back as the mid-nineteenth century, one has to wonder just how the concept of "deforestation" entered the discussion.[45] Other studies from the 1990s and the early twenty-first century confirm the general image of the Wallo landscape derived from the Maugini photographs, from rephotographing them, and from associated written sources, and bring more nuance to the discussion of landscape change and the potential roles of farmers in bringing it about.

Eastern Wallo Seen from Above

Aerial photographs and remote sensing are the most recent sources of evidence for landscape change in eastern Wallo. Belay Tegene, a physical geographer and a member of the MacArthur-funded collaborative research project, explored land use and land cover change in the Derekolli catchment area, more or less coextensive with the Gerba peasant association, drawing on fieldwork observations, interviews, aerial photographs, and remote sensing imagery.[46] Girmay Kassa wrote a master's thesis (directed by Belay) on changes in land use and land cover with special reference to soil degradation and population change in the Boru peasant association.[47] Kebrom Tekle and Lars Hedlund studied land cover change as evident from a comparison of 1958 and 1986 aerial photographs of Qallu *warada*, eastern Wallo. Qallu is immediately to the south of Tahuladare and presents the same ecological conditions and cultivation practices as found in Tahuladare.[48] Kombolcha is its principal town, and Mt. Yagof is immediately to its south. While the scale of the studies by Belay and Girmay are roughly comparable, the study by Kebrom and Hedlund covers an area more comparable to that of all Tahuladare.

Two factors complicate a comparison of the results. They cover differing time periods and, more important, use different categories of landscape classification. All start with the earliest available comprehensive aerial photograph, which was 1957 for Derekolli, 1958 for Qallu, and 1965 for Girmay, which in terms of political economy all fall within the period of Haile Selassie's post–World War II restoration. They all used the aerial photograph of 1986, which was taken at the height of the Darg period. Girmay and Belay included a third time point derived from remote sensing data—1994 for the former, 2000 for the latter—which allows for changes in political economy following the transfer of power from the Darg to the Ethiopian People's Revolutionary Democratic Front in 1991. The differences in time period, though bothersome, do not undermine comparison. The problem of land use categories is more complicated. Table 5.3 reduces their multiple categories to just three, highlighting the areas under cultivation and under

TABLE 5.3. Changes in land cover: Eastern Wallo, 1957–2000, percentage distribution

Source place, total area	*Types of land cover*	*1957, 1958*	*1965*	*1986*	*1994*	*2000*
Belay, Derekolli catchment, south Wallo, 1,527.9 ha	Cultivated	65.1		69.1		70.6
	Uncultivated vegetation	33.7		29.6		27.9
	Other	0.8		0.8		1.5
Girmay, Boru Metero, South Wallo, 3,089.1 ha	Cultivated		44.9	49.4	49.6	
	Tree farming		1.6	4.7	4.6	
	Uncultivated vegetation		35.1	29.5	30.4	
	Other		17.0	16.3	15.0	
Kebrom and Hedlund, Kalu District, south Wallo, 110.2 sq. km	Cultivated	49.6		50.8		
	Uncultivated vegetation	43.4		37.4		
	Other	7.0		11.8		

Sources: Belay, "Land-Cover/Land-Use Changes"; Kebrom and Hedlund, "Land Cover Changes Between 1958 and 1986"; Girmay, "GIS Based Analysis of Land Use/Land Cover." Girmay expresses his figures to the second decimal place. I have rounded to the first for the sake of comparison.

woody vegetation, and lays out the differences in categories as used by the different research projects.

For our purposes, the changes in the total area under cultivation and their relationships to gains or losses of woody vegetation are of greatest interest. The total area under cultivation changed the most in the *qolla* area of Gerba/Derekolli, where between 1957 and 2000 it grew from 65.1 percent to 70.6 percent of the total land surface, a difference of 5.5 percent. This is largest in terms of both total area and the extent of change. In 1958 and 1965 Qallu and *dagga* Boru had about half their area under cultivation (49.6 percent and 44.9 percent).[49] While Qallu saw a minimal expansion of the area under cultivation (from 49.6 percent to 50.8 percent, just 1.2

percent) in the years up to 1986, growth in land under cultivation in Boru grew up to 1994 at a rate approaching that of Gerba (from 44.9 percent to 49.6 percent, an increase of 4.7 percent). These numbers coincide in a very general way with the impressions gained from the matched terrestrial photo pairs, which suggest an area under cultivation of roughly 50 percent and only a modest increase in that area over the sixty years between 1937 and 1997.

These three studies of eastern Wallo are comparable to the 2001 study of landscape change in eastern Gojjam by Gete and Hurni because of the close similarity of their methodologies and the cultural and physical similarities of the sites.[50] The Gete and Hurni study area covered some 271 square kilometers and lay mostly within the boundaries of Dambacha *warada*, 48 kilometers northwest of Dabra Marqos, the principal town in the region. It consisted mostly of *dagga* and *wayna dagga* agro-ecological zones. They estimated population density at about 125 people per square kilometer. Their findings were alarmist. Comparing the landscape information in aerial photographs from 1957, 1982, and 1995, they calculated a loss of 99 percent of the area's natural forest, from 27 percent of the total area to 2 percent in 1982 and 0.3 percent in 1994. This is quite different from the Wallo studies that use the category of "forest." Girmay found 3.4 percent cover in 1957, declining to 2.6 percent in 1994, whereas Kebrom and Hedlund found 7.8 percent in 1958 declining to 5.4 percent in 1986. Gete and Hurni attribute a good deal of this forest loss to the expansion of the area under cultivation, which almost doubled, from 39 percent of the total area in 1957 to 70 percent in 1982 and 77 percent in 1995. Loss of natural forest was modestly offset by an increase in the area under afforestation from none at all to 2 percent in 1995. Table 5.4 summarizes their findings in a form comparable to those of the three studies of eastern Wallo.

The Gojjam study departs radically from the three Wallo studies, which firmly support the conclusion that from the 1950s to the 1980s and 1990s, expansion of the area under cultivation was modest and cannot by itself explain changes in vegetation cover. Belay concludes, "The popularly held view that expansion of cultivation is a primary force of vegetation clearing and environmental degradation was not found to hold true in the period considered as land-cover conversion to cropland was not a major phenomenon in the catchment."[51] Kebrom and Hedlund concur: "clearing of vegetation cover to gain additional cultivation areas was minimal."[52] Relative stability in the size of the area under cultivation in the face of rising

TABLE 5.4. Changes in land cover, Dambacha *warada*, eastern Gojjam, 1957–1995, as percent of total area

Types of land cover	*1957*	*1982*	*1995*
Cultivated	39.5	70.2	77.1
Plantations	0	0	1.9
Natural forest	27.1	1.7	0.3
Other uncultivated vegetation	32.9	27.0	18.7
Other	0.6	1.1	1.2

Source: Gete and Hurni, "Implications of Land Use and Land Cover Dynamics," 188, table 2.
Notes: Total area of the study = 271 km². Other uncultivated vegetation = grassland, temp. grassland, bushland, shrubland, grassland and bushland, and bare land. Other = bare land, towns.

population is apparently accounted for by the fact that by the 1950s practically all land that was feasible to plow was already in cultivation. The decreasing returns involved in cultivating ever steeper slopes had overtaken the value of labor involved in cultivating. This trend was reinforced by the government conservation measures, which I explored in chapter 4. These measures closed the hillsides to grazing or to further cultivation. The needs of an increasing population were met by more intensified use of existing cultivated land. The elimination of fallow was a major contributor to intensification.

Explaining the differences between the three Wallo studies and the Gojjam study is a challenge. A priori we have to allow for different parts of Ethiopia to have different trajectories of landscape change. This is obvious when we consider the rainforests of southwest Ethiopia or the arid grasslands of the east, but Wallo and Gojjam were not so strikingly different. They have similar environmental endowments, although Gojjam does get more (and more reliable) rain. Eastern Gojjam has been more closely integrated into the national economy as one of Addis Ababa's breadbaskets, whereas Wallo has been less fully integrated. Finally, the interpretation of aerial photographs and remote sensing data requires a good deal of judgment. It would be quite surprising for two studies to produce the same result. Given the standing of the scholars who produced these studies and their command of the technology, we are left with a divergence that will require additional research to reconcile.

By themselves the Wallo studies contain differences requiring some reconciling. In collapsing different categories into "uncultivated vegetation," in table 5.5 I eliminated nuance, which reflects peculiar characteristics of the

TABLE 5.5. Categories of land cover

Category	*Belay*	*Kebrom and Hedlund*	*Girmay*	*Gete and Hurni*
Uncultivated vegetation	Shrubland, shrubland grassland, grassland, valley rim vegatation	Shrubland, forests, riverine vegetation, remaining open areas	Bush and shrubland, grassland, mixed forest plantation, natural forest, woodlands	Other uncultivated vegetation
Other	Roads, town	Settlements (rural and urban), water body, flood plain	Swampy, water body, bare land	Grassland, temp. grassland, bushland, shrubland, grassland and bushland, bare land, towns

Sources: Belay, "Land-Cover/Land-Use Changes"; Kebrom and Hedlund, "Land Cover Changes between 1958 and 1986"; Girmay, "GIS Based Analysis of Land Use/Land Cover"; Gete and Hurni, "Implications of Land Use and Land Cover Dynamics."

different areas. Kebrom and Hedlund and Girmay accounted for water and seasonal changes in its distribution. The former take into account a small dammed lake and the seasonally fluctuating floodplain of the BorQanna River, while Girmay reported on Boru Meda and Sandag Meda, large natural grasslands subject to seasonal flooding, as well as on a variety of swampy areas. Neither of these categories appears to have been of much use to Belay, although "valley rim vegetation," similar to the "riverine vegetation" of Kebrom and Hedlund, was. Most difficult to integrate with the other studies are the "remaining open areas" of Kebrom and Hedlund. This category includes barren eroded land, mostly on mountain tops, open areas near homesteads and grazing areas. It is of more than casual significance since between 1958 and 1986 it grew from 3.9 percent of the total area to 16.9 percent, but on-site investigation did not help in interpreting indistinct information in the aerial photographs or allow for refinement within the category. Also distinctive and somewhat resistant to comparison are the different categories of woody vegetation Girmay admitted. These are five: tree farming, bush and shrub, mixed forest plantation, natural forest, and woodlands. I have given his "tree farming" special recognition in the comparative table, because of its relevance to chapter 5. His "bush and shrub" is readily comparable to categories used by the other researchers. The distinctions among the four remaining kinds of woody vegetation do not appear of particular relevance here.[53]

Clarity gained by reducing the categories of landscape analysis comes at the price of complexity. Kebrom and Hedlund pay particular attention to the shifts between their categories and demonstrate that change was far from unidirectional; rather, gains in one area might be lost in another. Table 5.6 shows how the modest change in area under cultivation in Qallu from 49.6 percent to 50.8 percent masked underlying dynamics, whereby almost a quarter (23 percent) of the area under cultivation in 1958 had been lost by 1986 to other categories of land cover ("remaining open areas," 7.4 percent; rural settlements, 5.3 percent; urban settlements, 4.1 percent; shrublands, 2.4 percent; and "others," 4.0 percent). At the same time the area under cultivation actually grew slightly (1.2 percent) at the expense of other areas (21.6 percent of shrubland; 14.4 percent of forests; 33.2 percent of rural settlements; 67 percent of riverine vegetation; 24.9 percent of "remaining open areas"; 28.0 percent of floodplain; and 4.9 percent of urban settlement).

All studies agree that shrub land declined in quantity and quality. The Maugini photographs support this judgment. The decline in shrub land relates directly to farmers' need for fuel, which Kebrom and Hedlund see as the most important force making for landscape change. Informants, whose views are more fully explored in chapter 6, claimed that shrubs had

TABLE 5.6. Changes in cultivated land in Qallu, 1958–1986, as percent of total land area

Land cover type	*Gained from*	*Lost to*
Shrubland	21.6	2.4
Forest	14.4	
Rural settlement	33.2	5.3
Riverine vegetation	67	
Floodplain	28.0	
Urban settlement	4.9	4.1
Remaining open areas		7.4
Others		4.0

Source: Kebrom and Hedlund, "Land Cover Changes between 1958 and 1986," 47, table 4.
Note: Table shows overall change from 49.6% to 50.8% of total surface area; 76.8% of the area under cultivation remained unchanged.

been an important source of fuel when they were young, but one that had declined in importance over the course of their lifetimes, to be replaced with a growing reliance on dung and eucalyptus. A few respondents reported that their parents had taken up planting this tree. More commonly, informants told of doing so themselves during the restored regime of Haile Selassie in the 1940s, 1950s, and 1960s, and others later still. Only Girmay addresses the presence of eucalyptus trees, which had become a major source of fuel, building material, and cash and by 1986 (if not well before) had become of significance to the landscape. In Boru, tree farms, consisting almost exclusively of eucalyptus, grew from 1.63 percent of the total area in 1965 to 4.71 percent in 1986, then declined marginally to 4.63 percent in 1994.[54] An overall audit of landscape change and its impact on woody resources needs to take into account how much the loss of shrubs has been offset by farmer tree planting.

Such an audit would have to address the vexing question of overall change in woody biomass on the landscape of eastern Wallo. In 1998 I provocatively suggested that the Maugini photographs raised the possibility of an overall increase in woody biomass because of the expansion of eucalyptus planting between 1937 and 1997. Belay has challenged this claim: "Crummey's contention that there was more woody biomass in the Welo region as a whole at present than in the 1930s has not been confirmed by this study or by that of Kebrom Tekle and Hedlund."[55] But it remains an open question at least worthy of further pursuit in light of differences in tree-planting strategies according to agro-ecological zone. Chapter 6 indicates that farmers have different strategies for tree planting according to agro-ecological zone, pursuing it with particular vigor in *dagga* areas like Boru, less vigorously in the *wayna dagga*, and tepidly at best in *dagga* areas like Gerba.[56] Belay's data are consonant with this general pattern. As we will see in chapter 6, farmers in Gerba do plant eucalyptus, but neither my landscape photographs of the area nor my memory suggest that this was to the same degree as in Boru or in the *wayna dagga* regions of Tahuladare.

If the changing dynamics between cultivation and woodland has been prominent in national and international discussions about landscape change and its contribution to famine vulnerability, changes in the availability of pasture is a question of land cover of equal importance to Wallo's farmers. Only Belay and Girmay recognize pasture as distinct from other forms of land cover. Kebrom and Hedlund collapse it into "remaining open areas." Belay found that in Gerba there was a steady increase in grassland and shrub-grassland from a combined 12.7 percent of overall land cover in 1957 to 21.8 percent in 2000 (see table 5.7). In Boru Girmay found a dramatic reduction in grassland from 10.3 percent to 5.1 percent between 1965 and 1986, but then almost a complete recovery to 9.9 percent in 1994. Apparently grassland is not the same thing as pasture. Wallo informants almost unanimously testify to the shrinking availability of pasture. The raw numbers mask qualitative questions about how palatable and nutritious changing grassland was and how suitable it may have been for the support of livestock. More important, the raw numbers obscure how much of the grassland may have been on slopes that were too steep to cultivate. Before 1974 hillsides were important for grazing, but since 1974 government conservation measures have closed those hillsides to grazing, without providing alternative sources of fodder. Conservation success has added to farmers' woes.

The aerial photographs allow us an overall view of changes in the Wallo landscape. They lack the texture and nuance of the terrestrial photographs but provide a useful framework for understanding processes at work in the former. The most extensive change in the area under cultivation is the increase of 5.5 percent in the *dagga* area of Gerba and the Derekolli basin, a confirmation of those who posit a downslope migration of cultivation. In other areas, expansion of cultivation was quite limited. The terrestrial photographs may be read in such a way as to confirm that dramatic expansion of area under cultivation did occur but that its occurrence was largely felt in the *dagga* areas. Only two of the three Wallo studies found the category of "forest" to be of use, while all spoke to various kinds of woody vegetation. Modest to begin with, it generally declined by a few percentage points. To be sure, the Wallo landscape was constantly changing, but over the period for which we have concrete evidence, those changes were modest and provide little support for apocalyptic views. Change was largely the work of farmers, and in chapter 6 I explore their world.

TABLE 5.7. Changes in grassland, as percent of total land area

		1957, 1965	*1986*	*1994, 2000*
Gerba (Derekolli)	Grassland	1.6	4.7	4.8
	Shrubland grassland	11.1	14.4	17.0
Boru	Grassland	10.3	5.1	9.9

Sources: Belay, "Land-Cover/Land-Use Changes," 14, table 4; Girmay, "GIS Based Analysis of Land Use/Land Cover," 83, table 15.

6

Farmers and Landscape Change in Wallo

FARMERS ARE IMMEDIATE AGENTS of landscape change in Ethiopia. Other forces are at work at different time scales—from the millennial effect of continental drift to the centennial effects of water and wind and climate to decadal effects of government and economy. At the level of lived experience, urbanization, the economy, and governmental action—taxation, road building, development and conservation policy, collectivization, and villagization—all play roles. While government and urban dwellers continue to look at Ethiopian farmers as agents of environmental destruction and development experts deem them lacking in essential knowledge, recent research into farming practices and photographic records of landscape change have suggested a more nuanced and complex story.

Chapter 5 reviewed the social science literature exploring famine vulnerability and the realities and potentials of Ethiopian rural society in light of the Darg's attempts to transform society and agricultural production in the years following the revolution of 1974–75. Rich as its findings were, the research left unanswered many questions about the relations between farmers and the environment, a task taken up by a collaborative research project titled "The African Environment: Experience and Control," funded by the John D. and Catherine T. MacArthur Foundation, which ran from 1992 to 1998. Coordinated through the Center for African Studies at the University of Illinois at Urbana-Champaign, four teams carried out research in four different African countries—Burkina Faso, Côte d'Ivoire, Ethiopia, and Mozambique—faculty and graduate students from Illinois working with counterparts in the African countries. The approach was multidisciplinary across the physical and social sciences and included anthropology, botany, climatology, economics, geography, geomorphology and soil science, history, and politics.

The Ethiopian team was coordinated by Dessalegn Rahmato, then at the Institute for Development Research of Addis Ababa University. Other members included Bahru Zewde of the History Department, Belay Tegene of the Geography Department, Sebsibe Demissew of the Botany Department, and the late Asnake Ali, a graduate student in history. My own work took place in cooperation with this research team, which chose the theme of the relationship between environmental change and famine vulnerability and opted to carry out its research in Tahuladare *warada* in the historic province of Wallo, implicated in the famine of 1973–74, epicenter of the famine of 1984–85, and site of earlier research by Dessalegn. The team chose to focus its work in three peasant associations in Tahuladare, representative of each of Ethiopia's major agro-ecological zones—Gerba for *dagga*, Gwobeya for *wayna dagga*, and Boru for *qolla*.[1] The whole team visited the area to pick the sites, each member followed up with individual research, and the team sponsored survey research in each of the peasant associations, carried out by locally recruited enumerators under the guidance of Sosina Asfaw, a graduate of Addis Ababa University and a graduate student at the University of Illinois. The information acquired from this survey is cited as "household survey." The team was also able to call on the expertise of Illinois professors Alex Winter-Nelson, Peter Lamb, and Donald Johnson.

The group reported its findings in a special issue of the *Journal of Ethiopian Studies*, in a collected book edited by Thomas Bassett and me, and in a number of articles.[2] The special issue of the *Journal of Ethiopian Studies* was the outcome of a conference marking the conclusion of the project in June 1997. The articles addressed the major environmental and conservation questions underpinning the government and NGO interventions directed to providing long-term answers to what were perceived as the environmental causes of the outbreak of famine in 1984. Were local farmers destroying the environment through their agricultural practices that caused irreversible and unsustainable soil erosion and permanent, ongoing loss of woody vegetation through continued expansion of the area under cultivation? Was deforestation a cause of drought? In answering the former question, the articles argued that the apocalyptic tone of public environmental discussions was unwarranted and that although there was clearly environmental stress, local farmers had strategies to ease that stress, strategies that would be essential components of larger interventions. While noting changes in tree cover, the articles questioned the term "deforestation" and delinked the issue from climate change.

Dessalegn demonstrated that contrary to what many government and NGO agents thought, environmental thinking had an indigenous history. He explored debates about conservation in the 1920s and demonstrated that the concerns of the 1980s had been anticipated by, among others, Fitawrari Mammo Seyoum, in effect the governor of Wallo in the 1960s. He noted Mammo's concern with afforestation and soil conservation, which Mammo saw as essential components of economic development. A major conclusion was that rather than local farmers, a larger contributor to "environmental degradation" was the government and the rights to land and environmental resources it promoted.

Dessalegn revisited environmental policy in his chapter in the book edited by Bassett and me. He summarized the chapter, which developed the thinking of his *Journal of Ethiopian Studies* article, in his own introductory précis:

> Anyone travelling in the early 1990s along the main road to Wollo and the north-east would have seen the landscape littered with the debris of the environmental rehabilitation program of the Derg. . . . They would have seen ruined stone terraces and check dams, with the stones strewn all over the slopes and now posing a great erosion hazard; disused bunds, many of which were partially demolished; and heavily disturbed woodlots and forests. . . . the cost of this program, which was implemented mainly in the second half of the 1970s and 1980s, may well exceed US$900 million, all of it provided by Western donor agencies.[3]

The explanation of this failure, Dessalegn argued, lay in the conflict between the country's national environmental discourse and what he calls the environmentalism of the peasant.

Sebsebe Demissew reported on a survey of the region's botanical resources. On the basis of field observations, he produced a comprehensive inventory of plant species, which he assigned to four different vegetation communities: dry afromontane forest and grassland, evergreen thickets, *Acacia* woodland and marshes around the area's two lakes, Hayq and Ardibo. He reported the testimony of elderly informants to the effect that there had been a marked decline in the number and size of indigenous trees, although apparently no loss of individual species. The process of decline could, he noted, on the basis of travelers' reports, be traced at least as far back as the early seventeenth century. He was unconvinced that such deforestation as had taken place was a plausible cause of drought.[4]

Belay Tegene addressed the soil knowledge and fertility management practices of farmers in the Nibo catchment basin in Boru peasant association. He found much to value in these practices, which were essential as the bases for successful interventions, although insufficient to resolve problems of erosion and the maintenance of fertility. Moreover, he came to a deeper understanding of the constraints which farmers face in trying to adopt chemical fertilizers and new techniques to control erosion.[5] Later research in Gwobeya reinforced Belay's earlier findings and led to recommended strategies for soil conservation methods based on preexisting farmer practices.[6] Belay also addressed the question of changes in land use and vegetative cover by using technology to analyze repeat aerial photographs from 1957 and 1986 and remote sensing data from 2000 in the Gerba peasant association.[7] He concluded that there had been considerable population growth, but cropland as a proportion of the total land surface increased by only 7 percent between 1957 and 1986 and imperceptibly between 1986 and 2000. Instead of breaking previously uncultivated land, farmers abandoned fallowing, which meant more intensive use of existing farm fields. Belay perceived a shrinkage in shrubland by 58 percent between 1957 and 1968 and its conversion primarily to shrub grassland and grassland. This he understood as a significant loss in woody biomass, which farmers used for fuel and to make charcoal, some of which they sold to the urban market.[8]

Bahru explored the history of forest management in Wallo, starting with an inventory of regional forest resources and paying special attention to Yagof State Forest, east of Dessie, the largest and richest forest in the area.[9] He argued that there had probably been a net gain in forested area, much of it due to planting exotic species at the expense of farming land. The loss of farmland and increasingly strict prohibition on the use of forest resources by local farmers had brought conflict between local people and state authorities, which exploded in 1991 during the overthrow of the Darg with local people cutting down a good deal of protected forestry.

My article in this project was a preliminary exploration of the nature of landscape change, which took as its point of departure the 1937 photographs of Armando Maugini, which I rephotographed in 1997.[10] I argued that the 1937 photographs showed little evidence of trees, whereas the 1997 photographs documented increases in the number of trees. There was reason to believe that farmers, far from cutting down trees, had been active in planting them. I reiterated these arguments in a chapter in the project's collected book, jointly written with Alex Winter-Nelson. We argued against the dominant Malthusian views, which held that "backward" technology and population growth had led directly to environmental degradation in the form of deforestation and that deforestation was a chief contributor to declining agricultural productivity and rising famine. Instead, our evidence led us to conclude that Ethiopian farmers had innovated and responded creatively to external change, not least with active tree planting. We argued that Ethiopian farmers were dynamic, rather than static; that they have adopted tree planting, rather than deforestation, as an environmental management strategy; and that changes in tree cover have little effect on famine vulnerability. To be sure, we did not minimize the strict constraints within which these initiatives occurred.

Some of the research reflected in the article coauthored with Winter-Nelson was carried out after the project's formal conclusion, when I returned to the area in 1997, supported by a research fellowship from the US Fulbright Program and between April and August, with the assistance of Merera Ejjeta interviewed elderly people in each of the project's three peasant associations and in neighboring areas documented in Maugini's landscape photographs. My goal was to learn something of the dynamics of landscape change through the experiences of farming people who had lived through the changes revealed by the photographs. I pursued this with extended, open-ended interviews with men and women, shaped by a set of questions, which asked about family history and changing access to resources as reflected in livestock, agricultural land, grazing, water, fuel, and cash. These interviews are cited as "elder interviews." Because deforestation had played such a large role in the narratives that sought to explain famine and that shaped relief and development efforts, I directed special attention to the informants' perceptions of landscape change as seen in changes in numbers and varieties of trees and their experiences as planters of trees and woody shrubs. Since population increase also shaped the availability of resources, further questions sought the informants' experiences of demographic change—the numbers of their siblings and children, their marriages, their education and the education of their children, and their geographic mobility.[11]

To place the Tahuladare information in some comparative perspective, I carried out additional interviews with farming men and women in two different *warada*—Armachachaho and Dambeya—in the vicinity of Gondar, some 500 kilometers to the northwest, in which I had previously carried out research on historic land tenure.[12] These *warada* share much with Tahuladare—language, cultural values, social structure and agricultural practices—but were largely untouched by famine. By studying them I hoped to isolate some of the factors contributing to famine vulnerability in places like Wallo.

The total informant pool consisted of 120 individuals. I had intended to interview equal numbers of men and women, but in the end I interviewed sixty-seven men and fifty-three women on an individual basis and carried out two group interviews with men chosen by local government agricultural development officers in Armachachaho and Dambeya. Eighty-eight individual informants (forty-nine men and thirty-nine women) lived in peasant associations in Wallo; the remaining thirty-two lived in Armachachaho (eight men and six women) and Dambeya (ten men and eight women). They came from the three main agro-ecological zones. Most of them (seventy-eight) lived in *wayna dagga*; twenty-seven lived in *dagga*; and fifteen lived in *qolla*. Their average age was 69.3 years, which meant they had been born on average in the late 1920s and reached adulthood in the 1940s. They had lived under four governments—imperial Ethiopian (1930–35, 1941–74), colonial Italian (1935–41), the Darg (1974–91), and since 1991 an ethnic federation.

Perceptions of Landscape Change

Informants experienced many changes over their lives. They grew up; they married; they had children, some of whom died and some of whom produced grandchildren; they moved about; they lived through two wars; and the world around them changed. Population increased, modern schools and medicine entered their lives, and familiar landscapes were altered. The role and scope of government changed; starting in the 1970s, it became more intrusive in their lives. Throughout these changes informants were active, often reacting and adapting, and sometimes creating change themselves. One field in which change was evident was the landscape.

The informants had been chosen, in the first instance, because they had lived in the landscapes appearing in the Maugini photographs and throughout the period covered by the matched sets of photographs, or because they lived in the peasant associations chosen by the MacArthur project. These peasant associations were all adjacent to Maugini-photographed landscapes, although I was unable to replicate the photographs he had taken of the *qolla* region from Gerba through Bati, a market site located at the edge of the lowlands. The interviews elicited informants' perceptions of landscape change. This was not a domain of innocence. The Tahuladare informants had all suffered the trauma of the famine of 1984. By the time I talked to them, farmers had been hectored for more than ten years about how devastating landscape change had caused the famine and how their recklessness in cutting down trees had caused the change. One informant, whom Bahru Zewde and I met in February 1997, told how in the aftermath of the famine he had been resettled in southern Ethiopia and how local people had greeted him—"So now you've come to chop down our trees, as well."[13] Realities of tree cover in 1937 to 1997 appear in the comparative photographs, at least for forty-five of the sixty-four Wallo informants (70 percent of those responding to the question), who judged that the number of trees had declined since their childhood. Conversely, a substantial minority, 30 percent, deemed that there were more trees now than before. Gender seems not to have influenced perception, since of the twenty female informants responding to the question, six (30 percent) thought there were more trees now. However, responses did vary according to agro-ecological zone. A plurality of informants in *dagga* Boru deemed there to be more trees now than earlier, whereas in the lowland peasant associations of Gerba and Chis Aba Lima, no informant thought this to be the case. Responses in the *wayna dagga* peasant associations of Gwobeya, Sulula, and Abisa fell between these extremes: seventeen of the thirty elder informants (55.7 percent) believed there to be fewer trees now, while twelve (40 percent) believed the opposite, and one (3.3 percent) believed there to have been no change in the overall number of trees.[14] Differing perceptions of changes in the number of trees may well have reflected farmer activities as private tree planters. I return to consider tree planting after a survey of their collective life experiences.

Farming Lives

The elder informants were all farmers, but not all had access to enough land and oxen to live by farming alone and were obliged to supplement their income in other ways. I did not interview any artisans, who play important roles in leather and iron working and in making pottery, but the informant pool included several children of artisans as well as traders, woodcutters, beer brewers, and day laborers.[15] Agricultural labor is highly

Figure 6.1. Qobbo village, Qorke peasant association
Source: Maugini (1937, *left*); Crummey (1997, *right*)

gendered—men operate the ox-drawn plows, sow seeds, plant, and thresh. Women run the household—managing the food, cooking, fetching fuel and water, tending home gardens, and caring for the youngest children. Women may assist men in preparing the fields for planting by following the plows and breaking up clods, by weeding, and by harvesting. Some of the informants were explicit that husbands and wives jointly made household decisions about what crops to plant and what investment opportunities to pursue. Surely they also made joint decisions about the education and marriage of their children.

Childhood

Farm work starts at an early age. A number of elders dated events to when they were old enough, perhaps four or five years old, to accompany their older brothers and sisters in pasturing the family's livestock. This chore was a heavy responsibility, but it provided children with the opportunity for play. Elders described other chores. For girls these were fetching water and gathering fuel, whether wood or dung from the fields. Boys also carried out these tasks, perhaps less frequently. My informants' generation received less formal education than did their children—20 percent of men went to school, whereas none of the women reported having had formal education. "Herding is our education," claimed FS, a woman in Gerba. For the men it was Quranic or church school.

One woman reported having learned embroidery, another had learned basket weaving, and several described spinning cotton and going to market. Girls might also carry meals to the men who were working in the fields. A couple of men reported having done wage labor for the Italians. Fighting was mentioned as a young man's activity, and one informant had been a servant for a prominent local man, whose gun he used to carry. While most elders' backgrounds were very similar, two women hinted at servile backgrounds, and one grew up in the household of a rich clergyman in the company of the children of servants (*ashkaroch*). She assured us she had experienced the usual chores, but had only done these as play, since the real work was done by the servants.

We did not learn much of the emotional experiences of childhood, although three elders in the *dagga* peasant association of Gerba recalled being caught up in the civil war that accompanied the efforts of the government in Addis Ababa to establish its authority in 1917 and that they and their families had fled before the invading Shawans. Another woman, AB, in Armachachaho, told us of her father, a patriotic resister to the Italian invasion. "I was very small and I couldn't know where he was going, but I wept when he left for his campaign," she said. While the majority of informants grew up in homes with their biological parents, a few referred to the breakdown of a household, through divorce and the experience of a step-parent, or the death of either or both parents. Where parental households had completely collapsed, informants turned to grandparents or to older, married siblings. This did not always work out when sibling rivalries came to the fore or when the sibling's spouse rejected the child. One informant, orphaned at an early age, spoke of growing up in the household of "friends."

Death was a companion from an early age. Of 120 elder informants, 88 (73 percent) reported the death of one or more siblings.[16] The average elder in Tahuladare belonged to a sibling set of about six, almost one-third of whom died before they reached adulthood.[17] Elder informants in Armachachaho and Dambeya had similar experiences. Here family sizes seem to have been a little larger, closer to eight children, and mortality lower, the average informant losing a fifth of their siblings, roughly two per set.

Marriage

Girls entered adulthood when they left home to be married; most boys left when they married and founded their own households, after starting to plow with their fathers. All elders had been married at least once, and about half of them were divorced at least once. On the other hand, just about half (forty-five of ninety-three reporting informants, 48.4 percent) never experienced divorce, suggesting that marriage was more stable than sometimes has been suggested.[18] Arranged marriages were normal and it appears that for many young women, the first assertion of their adult will came with their decisions to divorce their husbands. They would then return to their parental household, where in the arrangements for a second marriage they would have played more active roles. Men could also initiate divorce. One common reason for the breakdown of a marriage was the failure to have children.[19] Incompatibility—*ehelenna weha*, "like grain and water"—was an issue for both sexes.

Young women married at an earlier age than did young men. The average age at marriage across all informants was around twenty for men and twelve to fourteen for young women. The average age in Dambeya was lower than in either Tahuladare or Armachachaho—eighteen for boys and ten for girls. Dambeya custom provided for child marriage, something not reported by informants elsewhere. A number of Dambeya informants—men and women—reported five as the age at which some of the women had been married. To be sure, marriage was not the same thing as forming a conjugal unit, something that was supposed to await the girl reaching puberty.

Getting Around

Until very recently, the literature paid little attention to geographical movement as part of the regular rhythms of countryside life.[20] Most studies of migration explore Ethiopians moving from the country to towns and cities or responding to demographic pressure or famine-induced crisis.[21] Nevertheless, the 1984 national census paid attention to intraregional migration and noted that 80 percent of the movement within the Gondar and Wallo regions was rural to rural, suggesting that this kind of movement might be worth further study, a suggestion taken up by both Mesfin W. Mariam and Markos Ezra.[22] My elder informants indicated that movement was a real part of their lives, if constrained within the boundaries of adjacent peasant associations. Mesfin's findings were that his respondents (96 percent) were averse to moving elsewhere, an attitude possibly reinforced by the desire to avoid recruitment to the resettlement program. He also noted that most migration took place between places where people were related. Moreover, most of the reasons given for migrating had to do with family affairs—marriage, divorce, and the like.[23]

The patterns of movement sharply reflect the imbalance in opportunity between the genders. Although customary law among Amharic speakers provides for equal inheritance by sons and daughters, not least when it comes to agricultural land, in practice men, who alone guide the plows that start farming activities, exercise disproportionate control. As a result marriage tends to see women moving to their husbands' localities, and women moved more frequently and further than men did. Women left home to get married, often moving beyond the boundaries of their natal peasant association, whereas men more frequently continued to live closely connected to their original homes, farming the land of their fathers. Frequency of movement and a wider domestic circuit did not favor women when it came to long-distance travel. The longest trips were generally made by men, who much more frequently than the women moved to trade, find wage labor, and wage war.

Statistical information on geographical mobility reported by elders in Tahuladare, Armachachaho, and Dambeya tells part of the story. Of the male informants, 69.1 percent still lived in the peasant association in which they had been born, compared with 50 percent of female informants.[24] Men migrate across peasant association boundaries at a rate less than half that of women (22 percent versus 50 percent). At the lower level of the *got*, or village, men still lived where they had been born at a rate twice that for women (41.2 percent versus 21.2 percent). In contrast with male stability, over three-quarters (76.9 percent) of the women no longer lived in their natal *got*, whereas this was true for just under half of the men (48.5 percent). In their responses to the question of whether informants had ever lived anywhere other than their natal *got* and the *got* in which they were living when we interviewed them, the difference between men and women is equally marked. Less than half the men (45.6 percent) had ever lived any place other than where they were born and where they now live. By contrast, this was true of well over half (65.4 percent) of the women. Over their lifetimes women move considerably more than do men.

Of all the elder informants, among the women, BS had lived in the most other places, while ZM's movements had taken her over the widest territory. Among the men, YH of Qorke peasant association mentioned the most number of other places lived and the widest range of movement. BS was living in Gwobeya peasant association at the time of the interview. She had lived in three other peasant associations and in the towns of Hayq and Dessie. The distance from Dessie to Hayq is about twenty-eight kilometers. ZM was somewhat more wide-ranging. She was born on land where the Kombolcha airport was subsequently built. Her interview took place in another peasant association, beyond Passo Mille on the way to Wechale.

Her life had carried her about seventy kilometers from her natal home. Trade led YH over a vastly larger range. He resided in Qorke, but his trading had taken him to Gondar, Humera, Gamu, Gofa, Sidamo, Maji, and Assab, localities that included the lowlands on the Sudan border in the northwest, the rainforest of the southwest, the southern reaches of the Rift Valley, and the Red Sea coast. The area he covered was some 1,000 kilometers east to west and north to south. Traders were not evenly distributed across the three study areas, but were concentrated in Tahuladare, in the peasant associations of Qorke and Boru. Qorke, which is reported as part of the Sulula cluster, lies astride Highway 1, which runs from Addis Ababa to the Eritrean border. In contrast to Qorke, which has two crop seasons, Boru lies in the *dagga* agro-ecological zone and has only one, which limits agricultural activity but frees up time to invest in trade. Moreover, the *meda* of Boru, a large, grassy plain, provides grazing opportunities available to few other residents of *dagga* and *wayna dagga* Tahuladare and supports the horses and mules essential for trade. However, the kind of trade in which YH and the Boru informants had engaged is an opportunity largely lost to their children and grandchildren. Trucks and buses have displaced the horses and mules.

The children of elder informants move more and farther than elders did, although there is still striking local continuity. Regional capitals are magnets for the next generation—Dessie in the case of Tahuladare, Gondar for Dambeya and Armachachaho. Educational opportunity is one of the attractions of Gondar and Dessie.

Migration opportunities vary by region. Tahuladare is more affected by the out-migration of its children than is Dambeya or Armachachaho. Within Tahuladare, one peasant association stands out from the others. Gerba peasant association is on the old highway, which until the independence of Eritrea in 1993, linked Dessie (and before that also Addis Ababa) to the Red Sea port of Assab, some 750 kilometers away. This road is the principal route to the central Awash Valley with its commercial agricultural estates. A majority of the families in Gerba have sent their children—daughters greatly outnumbering sons—to the Awash and the Red Sea: eleven different families have sent nine daughters to Djibouti, four daughters to the Awash, and two sons to Saudi Arabia. Qorke, with weaker data, is a pale reflection of Gerba. From Qorke one daughter found her way to Assab, and two sons went to Assayta in the central Awash. Boru is also distinguished by the departure of its children, perhaps a legacy of the trading activities of their fathers. Six Boru families have sent children far afield: five daughters in Addis Ababa and one daughter each in Jimma, Awassa, and Jeddah, with sons in Jimma, Arba Mench, and Addis Ababa. In the generation of the informants' children, especially in eastern Tahuladare, women have greatly enlarged the ambit of their migration. This reflects the opportunities opened up by national highways. It probably also reflects the fact that for women, just as it was for their mothers, staying at home is less attractive than for their brothers, who have greater economic opportunities and perhaps greater control over domestic violence.

Collective social violence—vendetta and war—impinged more directly on men than it did on women, forcibly removing them from their homes and disrupting the lives of all family members. This information emerged freely in the course of the interviews, which did not explore the extent to which domestic violence was a factor contributing to the movement of women. Vendetta appears to have been of no importance in Tahuladare, where informants never mentioned it. It was a real factor in the lives of informants in Dambeya and especially Armachachaho. In Dambeya two of eighteen informants and in Armachachaho three of fourteen informants told stories of lives shaped and ended by vendetta. In Armachachaho AT told of how two of his brothers had died, one while searching for a lost ox, and the second, a priest, as part of the vendetta arising from the death of the first brother. ZA told of how one of his brothers, who had moved some distance away to escape a vendetta, was pursued and murdered. SA recounted how his brother had killed a man. A community investigation followed. The brother refused to confess and then, fearing vendetta, fled for refuge, only to be pursued and murdered. TT in Dambeya told of how his cousin, having murdered someone, fled to join TT; TZ told of how his family had relocated to where he now lived, his mother's country, for fear of vendetta.

Elders' lives spanned the decades from the 1910s to the end of the twentieth century. One notable period of peace (from 1945 to 1974) aside, warfare was all too common: wars to weld the national state in the 1910s and 1920s; wars surrounding the Italian invasion, occupation, and resistance; and wars unleashed by the outbreak of revolution in 1974. Informants experienced these wars directly: early childhood flight from civil war; involvement in the wars of the 1930s; and, after 1974, through the drafting of their brothers and sons.

The earliest experiences of war were those of the three Gerba elders, who, as very young children, probably in 1916, fled with their families before an army led by the Shawan general Habte Giyorgis, who was consolidating control by the Addis Ababa government over Wallo.[25] The father of another informant, KN of Dambeya, found himself a decade later in the last of the struggles to assert the power of Addis Ababa over the provinces of Wallo and Gondar. He was a soldier under Ras Gugsa, whose defeat in March 1930 ended the last holdout against the rising power of Haile Selassie.[26]

In October 1935 Italy invaded Ethiopia and occupied it until 1941, when a joint Anglo-Ethiopian force expelled it. Elder informants and their fathers were involved—on both sides in roughly equal numbers—in all phases of the war, which lasted throughout the period as a result of Ethiopian resistance. Four informants campaigned on behalf of the Italians, as did the fathers of three others, and a brother of another, while one informant and the fathers of four others fought in the Ethiopian army. TZ in Dambeya recounted how in 1935 Ethiopian government officials measured his testicles and, finding that they were the same size as his father's, drafted him. His father purchased TZ's release with 2 birr. HW, also in Dambeya, told how her father had been labeled a *banda*, the pejorative term for Ethiopians who served the Italians, because of local involvement with the Italian administration. By contrast, the second husband of FK, also from Dambeya, was a patriotic fighter.

The revolution of 1974 plunged the country into seventeen years of civil war, which only ended with the overthrow of the Darg by one of the insurgent forces. In its latest stages, the war came home to some Tahuladare informants—those living closest to the major highways along which battles and skirmishes occurred. An uneasy battle line ran through Armacachaho, Ethiopian People's Revolutionary Democratic Front (EPRDF) insurgents on one side, government forces on the other. I was conducting research in Armachachaho in 1988 and heard many stories of people paying taxes to both sides, since the tax receipt was the only witness to someone having recognized the political authority of either contesting party. One informant in Boru still had a bullet in her leg, which she had received from Darg soldiers, on her way home from market in Dessie. However, the wars had their harshest effect on informants across all the study areas with the Darg's institution of a military draft in 1983.[27] Hated at the time, it was commonly known in the countryside as *yalej geber*, "the tax of children." Fourteen families saw their sons drafted. Some returned, but we often heard the refrain, "and they were never heard from again." One informant, in Chis Aba Lima peasant association in Tahuladare, lost all four of her sons this way. Government and ruling party protestations of socialist equality did nothing to protect this woman, one of the poorest of the informants.

Farming Resources

The routines of country life depend on access to resources: water, food, and fuel for daily life; and land, seed, oxen, and grazing for farming.[28] Markets provide access to valuable commodities farmers cannot produce for themselves. They also provide opportunities for socializing, gathering and exchanging information, and expanding a farm's resources. But participation in the market requires that farmers have goods to sell or cash with which to buy them. Some cash may come as remittances from family members, but the rest must be earned from the sale of local resources.

Water, Food, and Fuel

We asked elders, particularly the women within whose domain it falls, about water. Fetching water is an arduous duty for women, but water is available and few alternatives to fetching it seem available.[29] Most farmers produce most of the food they eat, although there are no "subsistence" farmers in the study areas, if by that we mean farmers without significant engagement with the market. Some farms do not produce enough food to support the farm family, whose members must seek sources of revenue other than their crops. Other farms, which readily produce enough food to provide for the farm family, still turn to the market for food items they cannot produce—spices, sugar, and salt—as well as other household items like pots and pans, tea, matches, kerosene, and clothing.

Fuel has a place in the dominant environmental debate narrative. The assumption is that historically the principal source of fuel was naturally occurring wood. The story goes on to recount how a relentlessly increasing demand for fuel has led to an impoverishment of indigenous sources of wood. As wood grew in increasingly short supply, dung emerged as its principal alternative, and the use of dung for fuel means that it is not available as a fertilizer. We asked informants, especially women, about their use of fuel over time.

The trend in changing fuel use confirms, in a general way, the established narrative. There has indeed been a movement away from woody sources toward the use of dung and crop residues. Overall ninety-eight elders (95.1 percent of informants responding to this question) reported having used wood as a fuel in their youth; seventy-six (73.8 percent) reported doing so at the time of the interview. Informant use of dung as a fuel increased from forty-two (40.8 percent) in their youth to fifty-seven (55.3 percent) at the time of the interview. There was an increase in the use of sorghum stalks and crop residues from twelve (11.7 percent) to twenty-nine (28.2 percent) and a decrease in those claiming not to use dung from five to just one. Clearly dung has become increasingly used as a fuel, and there has been a decline in the availability of wood. However, the narrative overlooks the fact that 40.8 percent of 103 respondents report having used dung for fuel in their youth, and the measures respondents have taken to compensate for the decline in available uncultivated sources of wood. Fifty-two (53.1 percent) of 103 respondents have taken to planting trees. These trees are overwhelmingly, but not exclusively, eucalyptus; this has surely lessened pressure on uncultivated trees and on the use of dung as fuel.

There is modest regional variation in the changing sources of fuel. The trend toward increasing use of dung and crop residues is most apparent in Gerba, the *qolla* peasant association in Tahuladare, where the use of dung increased from 6.6 percent of elder informants to 73.3 percent and the use of crop residues from 6.6 percent to 80 percent. While the use of wood declined to eleven (73.3 percent) informants, this was not all that far off the regional decline to 67.6 percent. Gerba is the most lightly populated of the project's peasant associations. It has greater woody resources than the other peasant associations, mostly in the form of shrubs and bushes, and it was the only one where informants reported the sale of wood as a strategy for coping with famine. Confirming the trend, but very modestly, was Dambeya, where the use of wood for fuel declined from 94.1 percent to 82.4 percent and the use of dung increased from 52.9 percent to 64.7 percent. An anomalous development was in Gwobeya, where the use of dung as a fuel apparently declined from 66.6 percent of informants to 40 percent.

Access to Cash

Cash is now an essential part of country life. Elder informants needed cash for taxes and school fees, in addition to consumables and household wares. Few if any controlled much cash, but most had some. Their means of acquiring it showed the same kind of diligence and responsiveness to opportunity evident in other spheres of their lives. Their opportunities varied with their location.

Seasonal migration for labor is one common source of cash, but only six of my informants mentioned participating in it. The most widely available opportunities came from cultivating bush crops and eucalyptus trees, although with respect to bush crops, Armachachaho alone accounted for one third of the growers, Tahuladare for the remainder; with respect to eucalyptus, the great preponderance was produced in Tahuladare and of that, almost 40 percent was accounted for by Boru. Armachachaho, again, accounted for over half (60 percent) the marketing of fruits and vegetables. Livestock raising was particularly prominent in Gwobeya (42.1 percent of the Tahuladare total and 29.6 percent of the overall total), while Dambeya, by itself, accounted for almost half of the field crops. Gerba peasant association alone accounted for over half (57.1 percent) of the remittances. Opportunities for wage labor were modest and evenly divided across all three research areas. A global comparison of the opportunities to gain cash would suggest that farmers in Wallo make particular use of eucalyptus trees and remittances, whereas Armachachaho stands out for its production of bush crops and fruit and vegetables for the Gondar market and Dambeya is unusually prominent in the production of cereal field crops, reflecting its historical role as a regional bread basket.

Land

Most elders started farming with parental endowments. The fate of these endowments depended on the skills and hard work of farmers and, during the time of Haile Selassie, their skills at mastering the rules whereby landholding was continually adjusted.[30] The rules changed fundamentally in March 1975 with the nationalization of rural land, the imposition of strict limits on the holding of agricultural land, and the creation of peasant associations as the institutions within which land was controlled.[31] The outcome was a major redistribution, which resulted in a leveling down of individual holdings. In principle land was to be held on a basis of equality, varying primarily with the size of one's family. Renting land was now illegal, except for widows, who were allowed to rent out the land they could not work themselves. Just over twenty years later, two years after the proclamation of the Federal Democratic Republic of Ethiopia in 1995, there was a second major redistribution. While the legal framework of nationalization and the institutional framework of the peasant association remained in place, the rules changed again. To be sure, political patronage was at work in the Darg redistribution, but it was more salient under the EPRDF, which stripped land from anyone associated with the Darg and reassigned it to their clients, although none of the informants were left completely landless. The EPRDF land redistribution also addressed anomalies, which had developed in the two decades after 1975, and transferred land to the younger members of the peasant associations, who had found it difficult to create their own farms, and to single women—widows and divorcees—who had not fared well under the Darg.[32]

Only 3 or 4 elder informants started their adult lives landless. Most had inherited *rest* rights, although some informants in Chis Aba Lima had been tenants on *gult* land belonging to churches, or, in one case, to Empress Mannan, wife of Haile Selassie, and descendant of a powerful Wallo family. Of the 120 total informants, 76 provided information about their fate under the Darg's land redistribution program. Of these seventy-six, twenty-nine further reported on changes in their land-holding since the change of government in 1991, although at the time of the interviews only in Dambeya had there been a comprehensive land redistribution under the EPRDF.

In the Darg redistribution, four times as many informants lost land as gained it. A tenth of the informants saw no change in their land-holding. There seems to have been no difference according to gender in the Darg gains and losses. However, in Dambeya under the EPRDF the only person to gain land was a woman and 3 women experienced no change as opposed to only 1 man.

Project survey research of households in the three Tahuladare peasant associations revealed that all farms were very small, but there was a considerable difference in average farm size according to agro-ecological zone. In both absolute and per household member size, farmers in the *dagga* peasant association of Gerba enjoyed four times as much land as did farmers in *dagga* Boru and almost six times as much as farmers in *wayna dagga* Gwobeya. To some degree this discrepancy was offset in Gerba by lower rainfall and thus lower agricultural productivity, and in Gwobeya by two annual rainy seasons and two harvests a year. Within a peasant association, differences between households in land-holding were small but they were large enough to contribute to differences in the abilities of households to reproduce themselves and enjoy a modest degree of prosperity.[33]

Livestock

Livestock herds consist of cattle, sheep, goats, horses, donkeys, mules, and camels and can be large but were modest in most cases. A modest herd might consist of one or two oxen, maybe a cow or two, about five sheep or goats, and a donkey. A viable farm household needs two oxen to draw a plow. Of the eighty-seven elder informants who gave information concerning their fathers' cattle, only one, the son of a landless weaver, reported none, nine reported only one, and seventy-seven (88.5 percent) reported at least two. Of this seventy-seven, thirty-four (39.1 percent of all eight-seven) said that their fathers had more than two oxen, making them comparatively wealthy. This would make 88.5 percent of the households into which the informants were born to have been autonomous in agricultural production.

This is not the case today for young families developing their farming activities, although the situation is more complex than it might at first appear. The decline in the number of viable agricultural households is common in the contemporary literature and interviews did provide information in support of this view. Almost two-thirds of elder informants reported a decline in their cattle holdings, eighteen reported having only one ox, and seventeen reported that they had none at all. Households with only one ox are not fully autonomous, although they may still be marginally viable. Such a household can pool oxen with another, which allows both to plow. This is far from ideal because it takes time for oxen to learn to work together.

About two-thirds of informants reported declines in their holdings of small stock and equines. This information suggests an overall decline in the access to resources that all farming families have, although the strength of this possibility is mitigated by a lifetime cycle in household access to resources. Bauer writes of successive stages of expansion, dispersal, and replacement.[34] As a couple ages, it may divest its resources to its children, for whom establishing their own households is a desirable goal. Comparing the informants at the time of the interviews with their parents means comparing a household at the divesting stage of the cycle with a household (their childhood home) at the expanding stage. We should expect that in many cases informants would report fewer cattle than their parents owned. However, the decline here may go beyond what was expected. Whereas thirty-four of the eighty-seven informants reporting on parents' resources said that their parents were rich enough to own more than two oxen, only six of them reported this to be true of themselves. From one generation to the next, the number of households with only one ox more than doubled, from seven to eighteen. Unsurprisingly, only one informant unequivocally claimed that he had more oxen than his parents (three versus two). Interview information also suggests that even fifty years ago, Tahuladare farmers had fewer livestock resources than did their peers in Dambeya and Armachachaho. At that time only 27.3 percent (fifteen of fifty-five) of reporting

Tahuladare households had more than two oxen, whereas in Armachachaho 50 percent of informants made this claim and in Dambeya the proportion rose to 77.8 percent. The biggest herd in Tahuladare consisted of sixty to eight oxen, thirty cows, five goats, no sheep, five or six donkeys, and one mule, whereas eight of eighteen reporting households in Dambeya claimed herds as large or larger (the largest was twelve oxen and fifty to sixty sheep), and in Armachachaho three of the fourteen reporting households reported herds this large, the largest one consisting of eighteen oxen, forty to fifty goats, six donkeys, and one mule.[35] Markos rightly notes that holdings varied over the turbulent years of the famine.

Population density is one factor contributing to the difference in herd size between the three areas. According to the 1994 census, Tahuladare as a whole had 245 people per square kilometer, compared with Dambeya, with 180 people per square mile and Armachachaho with a mere 25. Gerba, with the largest herd in Tahuladare, is also the most lightly populated part of the Tahuladare study area. But population density is not the whole story, because Dambeya, over seven times more densely populated than Armachachaho, has more large herds than the latter.

Oxen are important because of their centrality to farming. Small stock are important primarily as capital—assets that can be sold off in good times for cash and in hard times for food.[36] Only thirty-seven elder informants provided information allowing for a direct comparison between their holdings of sheep and goats and the holdings of their parents. This is too few for firm conclusions. Nonetheless, the numbers are suggestive. Ten times as many informants had fewer small stock than or the same number as did their parents. Taken together with the numbers of livestock, this does suggest impoverishment.

Grazing

Grazing is essential for maintaining cattle and small stock. Everywhere the availability of grazing varies with the seasons, but in Gerba and in Dambeya the changing season provides the opportunity to send some of their stock to the lowlands. Farmers in Boru and to a lesser extent in Gwobeya reported access to local public meadowland (*meda*). Elder informants reported overwhelmingly that their parents had had no problem in gaining access to grazing. Twenty-two farmers reported no problem in former times, and only one reported there had been a problem. When speaking of their own access, the proportions were reversed, with twenty-two reporting problems today as opposed to three who felt their access was adequate.[37] The image of plentiful meadowland in the late 1920s and early 1930s is clear from the descriptions: "plenty," "plentiful," "no problem," and "wherever you wanted"; as are the limitations of the present: "very constricted," "difficult," "trouble," "very serious problem," and "none now." The explanation is not far away. To be sure, there has been the loss of some grazing to plowland, but more important has been a conservation measure introduced by the Darg and maintained by the current government: the enclosure of hillsides. This has the reasonable goal of arresting erosion and building up woody vegetation, but it was introduced with no thought for alternative strategies for supporting the cattle and small stock, whose grazing was excluded, nor for obtaining the fuel, grass, or medicinal plants farmers had hitherto drawn from the hills.[38]

Elder informants have developed alternative strategies for feeding their livestock, but not before reducing their numbers. The elder who reported that his father had had trouble with finding grazing also said that even then they had turned to buying grass from farmers with more land. Today a few households report pursuing the same strategy. Others resort to feeding their livestock with the residue of field crops such as sorghum stalks, a strategy that puts the need for grazing in direct competition with the needs for fuel and fertilizer. Gwobeya displayed the most varied grazing strategies. In addition to public *meda* land, a number of farmers reported "private" grazing plots and one, a remarkable innovator, described buying a commercial feed called *faranglo*, in addition to using weeds from his fields and a private grazing plot. Clearly, if population density is playing any role here it is the classical Boserupian push to innovation, since Gwobeya equally exhibited wider and more successful flexibility in producing for the market.[39] To be sure, Gwobeya farmers can produce two harvests a year, whereas those in Gerba and Boru are limited to one. Gwobeya farmers have the additional benefit of proximity to an important market in the town of Hayq.

Crops

The MacArthur-funded project generated one source of information about crops and cropping strategies through its household surveys in the three Tahuladare peasant associations of Gerba, Gwobeya, and Boru. The elder interviews generated another such source, which also covered other Tahuladare peasant associations as well as those in Armachachaho and Dambeya. We found that farms are exceedingly small. The average size of farms in

Gwobeya and Boru is very similar at about 0.1 hectare, whereas in lowland Gerba it appears to be five times larger, at 0.5 hectare. This does not result in a similar differential in production, since yields are lower in Gerba. The most constricted situation existed in Gwobeya, where a quarter of households surveyed had less than 0.05 hectare per person, while only 3 percent commanded more than 0.3 hectare per person. As small as such a difference may appear, it can have a significant impact on household viability. The largest area of any farm is dedicated to growing the cereals, which are the basis of the rural diet. Not all farms can produce enough cereal to meet household need. In *wayna dagga* sites in Wallo, such as Gwobeya, the cut-off point in farm size appears to be 0.15 hectare per person, farms with less being unable to meet immediate household needs.[40] Conversely, availability of at least small amounts of cereal to sell in the market would suggest surplus production. Of the farms surveyed, only 10 percent in Boru, 34 percent in Gerba, and 42 percent in Gwobeya reported being in this surplus position, meaning that 90 percent of Boru households, 66 percent of Gerba households, and 58 percent of Gwobeya households were food-deficit and had to meet needs for basic staples through market purchases. Not surprisingly, across the surveyed peasant associations these food-deficit households showed themselves highly sensitive to market prices, which they said affected their decisions about what crops to plant in farm plots.

Trade, once an important generator of cash, contributes much less today. Livestock sales are widespread, but the farm household survey indicated that the number of animals was small. The diversity of crop production as revealed by the household surveys in the three Tahuladare peasant associations is impressive. In addition to the usual cereals and the other major staple, pulses, farms also produce *chat*, *gesho*, coffee, vegetables, fruits, sugarcane, and trees. The table on specific farm activities shows that all households produced cereals, but it also shows that 83–97 percent reported production of tree products and 15–42 percent reported maintaining garden plots for market sale. This indicates a greater overall commercial value for garden plots than did the elder interviews, which suggested that only a little over 6 percent of households in Tahuladare and Dambeya maintained commercial gardens. However, these same interviews indicated that in Armachachaho 60 percent of respondents were producing vegetables and fruit for market sale and that marketing of cereal crops played an important role in Dambeya. Agro-ecological zone and proximity to market shape cropping decisions.

Farmer Tree Planting

Ethiopian farmers have a strong commitment to trees and bushy crops. Even in Gerba, where elder informants said that difficulties in obtaining water limited their ability to plant trees, the household survey indicated that 83 percent of households were doing just that. The interviews with elder informants suggested congruent results, that is, 99 of the total 120 (89.9 percent of informants actually providing information) planted trees. Only 10.9 percent did not. This proportion holds true when we consider just the Tahuladare elders, of whom eighty-two provided information on tree planting; seventy-three (89 percent of them) saying that they were tree planters. There were 201 respondents to the Tahuladare household surveys who provided information on tree-planting activities, and 82 Tahuladare elders mentioned it in the elder interviews. If we merge these sources of information, we have a total of 283 informants, of whom 256 or 90.5 percent claim to have planted trees. Tree planting, although ubiquitous, was modestly skewed according to agro-ecological zone. In *dagga* Boru, 100 percent of elder informants were tree planters; in *wayna dagga* Gwobeya, Sulula, and Chis Aba Lima so were thirty-nine of forty elder informants; in *qolla* Gerba and Chabisa the proportion of elder tree planters falls to 70.1 percent (nineteen of twenty-seven informants). The only point of divergence among these remarkably consistent results is the data on *dagga* farmers, where the household surveys identified 83 percent tree planters, in contrast with the elder informant interviews, which turned up 70.1 percent. Given that the elder informants were not chosen in a strictly random manner, the returns seem compatible.

The most commonly planted tree was eucalyptus, although others were also planted in significant numbers: *qulqwal* (the indigenous *Euphorbia candelabrum*), *qundo barabare* (the exogenous *Mentha piperita*), exogenous *Casuarina equisetifolia*, *wanza* (the indigenous *Cordia africana*), *grar* (the indigenous *Acacia abyssinia*), and *tid* (the indigenous *Juniperus procera*).

When it came to reasons for planting trees, both the elder interviews and the household surveys of the three peasant associations yielded essentially the same result.[41] Rank order of importance is the same—house construction, fuel, and sale—whether we confine ourselves to Tahuladare or take all three interview areas into account.

Household survey data were able to identify some of the factors leading to tree planting and some of the factors constraining it.[42] Farmers indicated different strategies in planting trees—some planted only a few around their houses; others planted trees scattered around their houses and their farm plots; still others converted farm fields into tree lots. Tree lots were usually eucalyptus but also included indigenous juniper and acacia. The most obvious factor affecting strategies has already been identified—agro-ecological zone. *Dagga* areas are wetter, higher, and more densely populated, and their farmers are the most likely to plant trees in lots. *Qolla* areas are warmer, drier, and least densely populated, and their farmers are least likely to plant trees.

Statistical analysis indicated that wealthier farmers—those with large animals, food surpluses, formal education, and market gardens—are more likely to invest in tree lots than are poorer farmers, as indicated by the absence of an adult male in the household and reliance on small stock. Households with nonfarm income are less likely to invest in tree lots, suggesting either that these activities drain labor and other resources away from the household or that nonfarm income is an indicator of poverty. Neither knowledge nor marginal changes in population pressure seem to affect the decision to invest in tree lots. Although land availability does not seem to be a positive incentive for farmers to plant tree lots, unavailability is indicative of decisions not to plant tree lots. Overall the data indicate widespread tree-planting over a long period of time. The main barrier is relative poverty within a peasant association context.

The strategies of poorer-off households are highlighted by a survey which covered all of north-east Ethiopia, including eastern Tigray together with north and south Wallo. The survey found that 50 percent of poorer villages reported income from casual employment, 31 percent from firewood sales, 26 percent from laboring in towns, and 18 percent reported temporary migration in search of agricultural labor. Small numbers of poorer villagers also reported income from handicrafts and sale of cash crops and such households participated in food-for-work schemes at a higher rate than did their better off neighbors.[43]

The histories told by elder informants bring some of this information to life. The most remarkable tree planter interviewed was KG in Gwobeya peasant association. He was born around 1929, and when he was about twelve, after the Italians left, he learned to read and write. Originally from Kutabar, which is about twenty kilometers from Dessie, he moved first to Boru, where eventually he entered the employment of US missionaries running a hospital at Boru Meda. After six years he moved again, to Bestima, east of Hayq, perhaps a somewhat longer distance away. He had relatives there. He opened a tea house and produced and sold mead (*taj*), making enough money to send his children to school. He obtained land on contract and produced vegetables, which he sold in the market at Hayq. Meanwhile, he invested in farmland closer to Hayq in the vicinity of Gwobeya, where he moved around 1970. After the revolution broke out in 1974, he acquired land on the same basis as other members of the newly formed Gwobeya peasant association. He did not start planting trees until his move to Gwobeya, at which point he became very active and planted more than ten different kinds of trees and bushes. He sells oranges, *chat*, potatoes, and onions. In the interview he described how he experimented with trees, planting anything that caught his fancy. Neither landlessness nor Darg policies proved an obstacle to his farming ambitions.

Landlessness was equally no barrier to KB's ambitions. We interviewed him at the locality of Gaddiso, adjacent to Boru Meda. His mother's divorce and remarriage deprived him of land rights he might have inherited through his father. He channeled his energies into trading, the profits from which he invested in agricultural land. Revolutionary land redistribution after 1974 let him acquire enough land to give up trading and concentrate on farming. He had already started to plant trees in the late 1940s or early 1950s on land he had bought from local landholders. One of these plots he planted entirely in eucalyptus, which he used for house construction and to sell.

We have already seen that tree planting by farmers was not confined to Wallo but was also practiced in Dambeya and Armachachaho. In Armachachaho farmers planted at least seventeen different species of trees, including three species of fruit trees, which they planted with marketing in mind. They also grew bushes. While *chat* was the most popular bush crop in Wallo, in Armachachaho it was a poor third, behind *gesho* and coffee. One Armachachaho elder, MT, told of planting twelve varieties of trees, which provided the bulk of his livelihood, since he sold tree crops to buy food. Some of the trees he planted to shade his coffee bushes. Although less prominent than in either Wallo or Armachachaho, tree planting was also practiced by a strong majority of Dambeya farmers, although barely half of them planted bushes and none of them planted fruit trees.

Farmers and Landscape Change

Ethiopian farmers shape the rural landscape. Tree planting is one way they are doing it. This is not a practice that emerged out of revolutionary tree-planting schemes of the late 1970s and 1980s or from famine relief and rehabilitation projects of the 1980s and early 1990s. Government and external agency initiatives played small roles, although some elders spoke of how, in the days of the monarchy, local notables and governors were early planters and some of them encouraged others to plant. The role of Menilek's government in introducing the eucalyptus tree to Ethiopia is well known,[44] but none of the elders interviewed attributed their first knowledge of this tree to a government agency. Eucalyptus was well established in urban centers like Dessie and in lesser regional centers.[45] It was known in the countryside of Wallo and Dambeya before the Italian invasion of 1935, but it had not been widely adopted there. Elders spoke of first seeing eucalyptus as a planting of their parents or in a regional center like Boru, Bati, or Dessie. Informants in Gerba told how they had got their seedlings of eucalyptus and other trees, from Tamalso Kassaye, an Eritrean farmer and hotel owner in Bati. Others spoke of how they had developed their own nurseries. Still others obtained seedlings from government nurseries, first established under Haile Selassie and expanded under the Darg. The adoption of eucalyptus appears to have been a slow, informal, local process—neighbor to neighbor, relative to relative.

Farmer tree planting in northern Ethiopia started several decades before the transformative events of the 1970s and 1980s. Research also addressed the information provided by ninety elder informants about the beginning of their tree-planting activities. Five informants (four in Wallo, one in Dambeya) had already started planting when they were young, either just before or during the Italian occupation. Other informants dated their tree planting to life events—when they got married, the birth of a particular child—or public events, which could be traced to a particular decade. Others simply point to the reign of Haile Selassie, which in this context ran from 1941 to 1974. Altogether three quarters of the informants (sixty-seven of ninety or 74.4 percent) providing information on the matter started planting trees between 1941 and 1974, which, added to the five who had started earlier, brings the total to seventy-two or 80.0 percent of informants planting trees before the advent of the Darg in 1974. Finally, eighteen informants (20 percent) sad they had started planting trees between 1974 and 1991. This is also of significance because much of the literature maintains that Darg land policy strongly discouraged planting trees.

In imposing their policy of land nationalization and redistribution in the later half of the 1970s, the Darg nationalized tree lots, using land tenure as a conservation measure. The Darg was very active in large-scale tree planting, imposing trees on the enclosed hillsides and watersheds. Numerous informants told how they had participated in these plantings, sometimes as part of food for work schemes, sometimes because of coercion. When asked the separate question about whether the Darg had encouraged private tree planting, our elder informants responded more affirmatively than we expected. Only in Sulula did we get a strong negative response to this question, although even here more informants (six) said that the Darg had encouraged private tree planting than held the opposite view (five). Reconciling the prevailing view with the responses of our informants is not easy. One line of reconciliation would suggest that informant planting excluded woodlots and involved only trees around their houses and scattered across their property.

A more perplexing question is how farmer initiatives in general and commitment to woody crops in particular have gone so unnoticed and how Malthusian rhetoric has acquired such a hold on explanations of landscape change. Governmental and elite commitments to global narratives of deforestation and desertification have obscured what is happening in real life. If one fails to ask the question, one cannot expect to get the answer.

APPENDIX

Life Histories

Life History of a Wallo Farmer

Warqnash Tafarra

We interviewed Warqnash Tafarra on May 21, 1997, in the village of Qobbo in Qorke peasant association in Wallo. Warqnash was a child of about four years old when the Italians came in 1935, so she was about sixty-five in 1997. She was born in a nearby peasant association, one of five sisters. She had no brothers.

As a child, she collected firewood. In those days the main fuel they used was acacia wood. She also gathered firewood from the *agam* (*Carissa edulis*) and rose bushes. "We had no trouble in finding firewood in those days," she said. They did not collect dung, something she started to do after she moved to Qobbo, near the eastern escarpment. There were no eucalyptus trees at her father's home. She first saw them at a nearby town.

She came to Qobbo when she married Yusef Yemar, who died five years ago. She was married the year after the Italians left, when she was seven years old. There were acacia and juniper trees when she came here, also a lot of *agam* and rose bushes here, but no eucalyptus and no euphorbias. There were some olive trees, but no *wanza* (*Cordia abyssinica*), which could be found at Kombolcha, a town near the escarpment southeast of Dessie. She and her husband brought their first eucalyptus seedlings from nearby Qobbo.

Altogether she had eight children, three of them before she and her husband started to plant eucalyptus, which would have been in the early 1950s. This was something she and her husband discussed and decided to plant it "to deal with our problems": to buy clothes and grain and pay taxes. Originally they had not planned to sell it but to use it for house construction and firewood. Eventually they did sell it, but it never contributed greatly to their income, which she also augmented by selling chickens from time to time. They planted eucalyptus on both sides of the main highway from Dessie to Maqale and Asmara, which runs just above Qobbo. Some of the land was inherited and had been planted in grain earlier. In 1997 she still had this land, and it was still planted in trees. However, when I revisited her in 2009, she had lost the plot to a road-widening project. She felt that she had been inadequately compensated for the loss of this resource, which she considered her "pension." The only other tree she and Yusef ever planted was euphorbia. They never planted any cash-producing bushes.

We asked her about what role governments had played in her tree-planting activities. She told us that in Haile Selassie's time people planted trees on their own initiative. They were never encouraged to plant trees, although they were required to fence the compound of the local governor and some administrators. The Darg ordered people to participate in public planting schemes, and she planted in many places without compensation. The Darg did not encourage people to plant trees for themselves. In many ways the Darg period was a good one for farmers, but "they took our children" (probably a reference to the military draft, instituted in the 1980s). The Ethiopian People's Revolutionary Democratic Front government, which came in 1991, has never told them to plant trees. So far as government in general goes, her attitude is "Let the government be whatever it wants to be: I am laboring for myself."

When famine came in 1977 Eth. Cal. they got food relief in the form of wheat at the nearby town of Hayq. One of her daughters was resettled in Wallaga.

As far as resources go, years ago she had more livestock—two oxen, five sheep, one donkey, one mule—than she does now. They had enough private land to graze the animals on. Some of the livestock died; others were sold.

Of Warqnash's eight children, one son died after completing eleventh grade; one son is in Addis Ababa; and one son is in Assayta, a town in the Afar lowlands, many miles to the east.

I was able to revisit Warqnash in 2009. In this she was unique among my informants. She had lost her tree plots to a road-widening project. She felt that she had been inadequately compensated for the loss of this resource, which she considered to be her "pension."

Mt. Yagof as seen from Qobbo

Life History of a Wallo Farmer

Abagaz Adam

We interviewed Abagaz Adam in the village of Gaddiso, which is in the Boru peasant association. It lies in the *dagga* agro-climatic zone. Abagaz was born here. He was about eleven years old when the Italians came, so at the time of the interview he was seventy.

His father was a farmer, a horse trader, and an excellent horseman. He fattened horses for sale to the Italians. He fed the horses by cutting *katama* (*Schefflera abyssinica*) and *asandawa* (*Echinochloa crus-galii*) grass from nearby Boru Meda and bringing it back to feed the horses. He kept many mares for breeding but never dealt in mules. Abagaz's father was a tenant (*chisenya*) on church land and paid one-third of his harvest to the landlord, Mameé Alaqa Awaze, although he had inalienable hereditary rights to the land (*aynaqlagna*, *restennya*). His father was a representative for the church, and their landlord lived near Warra Ilu, a long way away. They had other dues on their land: they paid annually ten *berr*; with their neighbors they plowed and harvested church land; and on the church's annual patron feast day in Hamle (July/August) they would grind four *qunna* of barley to make beer.

Farming and Trading

As a boy he farmed and fattened oxen to sell in the Raya District of Tigray. He started Quran school but had to quit because there was no one else to do chores.

He married in 1939 at the age of twenty-two. Two years before that, when the militia (*nach lebash*) was formed, he went to Jimma to sell mules. Again in 1946 he went to Jimma Agaro to collect coffee. He also went to Teppi in Illubabor to fetch coffee. While he was trading, his younger brother farmed for him, and they shared the proceeds of the trading.

As a young farmer he grew barley, wheat, oats, beans, and peas. Later maize came; during the Darg days Chinese maize and a new kind of wheat came. "And now there is yet another new kind of wheat, which we are compelled to take along with fertilizer and for which we must pay."

With his profits from the trade, he bought clothes, grain, oxen, and riding horses, but he did not pick up any new ideas about farming. There was the coffee, of course, but it would not grow here.

Landscape

One of the most dramatic landscape changes in Abagaz's life has been the change in Boru Meda. When he was a boy the *meda* was much marshier than it is today, too wet to allow sheep to breed there. Moreover, there was a kind of grass called *gudin* that flourished then. If the sheep ate it, their throats would swell and they would die. That grass is still around, but the *meda* dried up after the Italians came.

Trees

When Abagaz was a boy, there were many juniper trees up on the hill, but there were no olive trees. The juniper areas became grazing land. People cut the junipers, which do not rejuvenate the way eucalyptus does. The Italians cut a lot of junipers, even from the cemetery. Later, under the Darg, they cut down juniper trees to preserve their land claims. The Italians gave no encouragement to private tree planting, although they did plant eucalyptus and *Shushuwa* along the roads. Under Haile Selassie, rich people planted eucalyptus. For example, Makonnen Lamlam planted eucalyptus at Wayramba and Amora. He stated that the Darg did not encourage private planting, although they organized large-scale tree planting. Eucalyptus does have disadvantages. It does not help control erosion, although it may contribute to climate improvement. The best measure against erosion is terracing.

Abagaz first planted about fifty eucalyptus trees during Haile Selassie's time. He spent about eight years at the American hospital, and when he returned he cut the trees to build a house. He also sold some to buy grain. He has built two houses with wood from his tree lot. This lot was previously grassland, but everybody else planted trees around it, so he followed suit.

Overall, there are more trees now. The whole land is covered with trees.

Today for fuel, he uses more dung than wood. As for fuel wood, in the old days it was juniper and acacia; today it is eucalyptus.

Family

Abagaz was one of seven children, four boys and three girls. Three have died (two boys and one girl), two after reaching maturity. Both of his remaining brothers were farmers and traders, and their trading took them to Jimma. The three sisters married at Dessie, Esmamo, and near Shawa Robit. Abagaz himself had five children, two boys and three girls. One daughter works for the Ethiopian Electric Light and Power Authority in Awassa; one daughter is married in Jimma Agaro; and one is married at Abarabo in Sost Gadal. One of his sons is in grade 1 in a government school, and the other son was drafted by the Darg and never returned.

Gadisso (Boru area)

Life History of a Wallo Farmer

Laqqach Kassa Tigray

We interviewed Laqqach at the village of Walkaganda in Aba Lima peasant association. Although she never said so directly, it is clear that she has struggled for most of her life, more so than many of her neighbors, who are far from privileged.

Laqqach was born in Waldalulu about five years before the Italians came, so she was about sixty-six when we interviewed her. Her father was a Tigray trader who settled in Waldalulu, where he was a tenant of Dajjazmach Gabra Egzi'abeher. Her mother, Zanabach, was also from Tigray, but Laqqach does not know the name of her mother's father. Her mother was simply called Zanabach Tigray.

Her father had two cows and three or four calves, which he grazed on the *meda*. He used the landlord's oxen to plow on a fifty/fifty sharecropping basis.

Laqqach married locally. Her husband, Hayla Mikael Kidane, farmed as a servant (*ashkar*) for a member of the *nach lebash* militia called Kabbada Getaw. Hayla Mikael quarreled with Kabbada Getaw and went to the house of Laqqach's father, but he quarreled there, too, because his father-in-law would not give them any land.

They moved to this general area, where Hayla Mikael started to farm for a new lord, Walda Mikael Asfaw, and Laqqach joined him a year later. Walda Mikael Asfaw lived at Qanagna kundu, "up there," and they lived as his servants. For a while Laqqach left her children with her father and served as a maid—grinding, fetching water, and fuel. She brought the water from the Qagna Weha River, which was far, maybe a half an hour away. The fuel was wood from the lord's woodlot, where there was *agam* (*Carissa edulis*) and acacia. They burned only wood, and not dung. Sometimes she went to market. Her husband took care of mules and guns for the lord. He got land as wages and they farmed sharecropping on a fifty/fifty basis with people who had oxen. They would leave the crop residues in the field for owners of the oxen. She got food at the lord's house, but they had to clothe themselves with the income from their fields.

After five years they moved to land belonging to the wife of the militia-man from Waldalulu, who had originally suggested they come to this area. It was administered by the woman's brother. Laqqach and her husband took care of coffee trees and farmed *tef* and maize by hoeing the ground. They grew about ten *qunna* of *tef*, maybe less. The lady gave them this land, which was up on a steep slope. Monkeys and baboons plundered the crop. They stayed there about two years.

Then they moved to the town of Wechale, leaving everything behind. At Wechale they worked at weeding fields for the local farmers, for which they got maybe one to two birr a day in cash. Overall, her husband may have made two to three birr a day, since he also farmed for other people for wages. They rented a house in the town for five birr a month, and all their income went for rent and clothes.

After the Darg came to power, Laqqach and her husband came to Walkaganda, where they got land from the peasant association. Although it can bring its own problems, owning your own land is better than not, she stated. After coming to Walkaganda they experienced villagization, which took place near the school, at Bageddo. She lived there for four or five years before returning here.

In 1977, to cope with the famine, they sold wood from their trees in the market and later got relief from the Mekana Iyasus mission. Hayla Mikael, who had been sick, did not survive the trouble.

Laqqach considers both her old country and this new country to be *wayna dagga*. In Waldalulu they had grown *tef*, maize, sorghum, beans, and another kind of sorghum called *zengada*. Here they grew *zengada*, barley, beans, and peas. Laqqach now grows cabbages, maize, coffee, and trees. She sells fruit of the *fula* on an annual basis and uses her wood for fuel. She also has a few eucalyptus trees for house construction. Even today the family farms by hoeing because it has no livestock. Their land down on the *meda* at Abbadera is farmed on a fifty/fifty sharecropping basis by someone who owns his own oxen. He sows *tef* and sorghum.

As a girl Laqqach collected fuel. Since Waldalulu is *meda* country, there were no trees there. For firewood they cut the eucalyptus, which grew in their compound, and used dung for fuel. Wood was more important. Since coming here Laqqach has burned leaves, *dadaho* (*Eucla schimperi*), and dung, when she can find it. When she came here, there were bushes such as acacia, *agam* (*Carissa edulis*), and *qantaffa* (*Pterolobium stellatum*), but no trees.

Family

Laqqach was one of ten children, three boys and seven girls, of whom four (three boys and one girl) died young. Two of her sisters married Italians in Dessie and went to Italy. One sister married in Kombolcha. She does not remember the names of her other sisters. Laqqach and Hayla Mikael had six children—four boys and two girls. Five had already been born by the time they left Waldalulu, and another one was born after they came here. One of the daughters married in Warra Ilu and later, again, in Addis Ababa and Dessie. This daughter does not work. The other daughter married a schoolteacher at Goraganda; she is deaf and dumb. She is a skilled hair dresser. Laqqach's son-in-law and all four of her boys were conscripted by the Darg, and none have returned or been heard of since.

Chis Ababa Lima landscape

Life History of a Wallo Farmer

KH

We interviewed KH in the village of G. in Gwobeya peasant association. Gwobeya was chosen as the *wayna dagga* site for the MacArthur collaborative research project. KH was a man by any standard—an innovator, experimenter, and tree planter. By local standards he had ranged fairly widely.

He was born at Allada in Kutabar, which lay on the other side of Boru. He was a boy taking care of his father's cattle when the Italians came, so at the time of the interview he was about sixty-eight years old. His father farmed *zamacha maret*, land that was under the obligation of providing military service. KH understood that form of tenure to be *rest*—under heritable right. As his military obligation, his father campaigned to "Asmara." KH's father was a prosperous man and owned two oxen, two cows, three calves, fifteen sheep, six goats, one donkey, and one horse. The goats browsed on the steeper slopes, while the sheep grazed on the *meda*, and the cattle on two separate private grounds and also on the public land at Hitalbo.

For fuel, dung was more important to them than wood, although they also had plenty of that: juniper, eucalyptus, *kosso* (*Hagenia abyssinica*), *embwaso* (*Rumez nervosus*), and *tsarsarsa*. Already in Italian times there was eucalyptus at Allada.

KH started church school at Kutabar in 1935 Eth. Cal., then attended a government school, where he was promoted to the fourth-grade level. He left after a year, but he had learned to read and write. He left school to farm. His land was small and still under the obligation of providing military service, which he discharged by doing guard duty at the *meslane*'s office in the countryside. In 1948 he moved to Boru Meda and worked for the *faranj*, at the American mission hospital there. He cooked and was chief of the gardeners. He learned how to make food by writing down the instructions. He stayed there six years, earning 40 birr a month. He lived on the compound and had his own gardener but felt that his salary was not enough, so he moved to Bestima, where he had two aunts and friends. At Bestima he opened a tea house and *taj bet*.

He stayed at Bestima for about fourteen years and made enough money to send his children to school. He obtained land on contract and grew vegetables—*tekel goman* (a variety of cabbage, *Brassica oleracea var. capitata*), carrots, and potatoes—and *tef*. He sold these vegetables in the market at Hayq. During this period he bought land at Gwobeya for about 200 birr.

KH moved to Gwobeya in 1959 Eth. Cal. He had quarreled with someone in Bestima and feared their revenge. He then bought land in four different places from four different people and farmed *tef* and vegetables for four years before the Darg came to power. He got land under the Darg. In 1972 Eth. Cal. a producers' cooperative was formed. KH was a founding member and later chairman, but he found the experience irksome—people just did not cooperate. In 1978 Eth. Cal. he moved to the place where we interviewed him. The cooperative was dissolved in the last year of the Darg and its remaining assets divided among the remaining members.

Famine

The drought made 1977 a year of great trouble.

Tree Planting

KH did not plant trees at either Boru or in Bestima, but he has been very active planting trees in G. He has more than ten kinds, all of which he planted—eucalyptus, acacia, *shushuwa* (*Sesbania sesban*), *tini* (*Opuntia vulgaris*), *besana* (*Croton macrostachya*), foreign juniper, *ensät*, *agalo* (*Combretum molle*), euphorbia, avocado, and orange. He has 5 orange trees, 6 avocado, 12 coffee bushes, 90 *chat* bushes, about 150 eucalyptus, 1 *wanza* (*Cordia africana*), and 1 *bisana*. "I like trees," he told us. He sells oranges, *chat*, potatoes, and onions. Potatoes sell for about 90 birr a quintal and onions for 100 birr a quintal. In a year he might sell eight quintals of onions and ten quintals of potatoes.

Today KH has three oxen, two cows, three calves, four sheep, and one donkey. For his livestock he buys a feed called *farangalo*, which is made from the residue of vegetable oil processing and costs about fifty birr a quintal. He also buys grass. "This year," he told us, "I paid 500 birr to feed my cattle."

Family

KH's father had ten children (four boys and six girls), of whom three girls died young. Of his adult sisters, one lives in Addis Ababa, where she works in a *buna bet* (coffee house); one lives at Ruga, near Kutabar; and one lives at Bruktu, beyond Alasha. All three of his brothers are farming at Allada, where KH had been born.

KH divorced and then remarried his first wife. He later married another woman in Bestima. His first wife had five children: one boy and four girls, of whom one girl died. His second wife had two boys. One daughter lives in Dessie, where she works for UNICEF and is married to a government worker. Another daughter, who had originally married in Hayq and then spent a long time in Jeddah, also now lives in Dessie. A third daughter attended typing school and is not married; the fourth also lives in Dessie.

KH's 2 sons are not married. One lives with his mother in Jimma and the other lives at Awasharba.

Gwobeya peasant association

Life History of a Wallo Farmer

ASA

The interview with ASA took place in K location in Gwobeya peasant association, in the *wayna dagga* agro-climatic zone. She had been born here, nine years before the Italians came, so she was about seventy years old. She was first married during the third year of Italian rule. Her father was a *qalecha* (soothsayer) and a *gabbar* (farmer). As *qalecha* he did not preach; he led prayers.

ASA lived for a time on a piece of her father's land. Her family never drank alcohol, since they were descended from a line of *abagar* (Muslim holy men). Her mother was Nurit Muhe, and her mother's father, Ibrahim Sharo, held this title, which has passed to his descendants.

ASA grew up in the house of her elder sister because her mother had died when the Italians came. As a girl she tended cattle, weeded, fetched water from Hayq, and gathered firewood. They used to wash their clothes in Hayq but stopped doing so when a fishery was developed. They now get piped water, across the road, from a spring called Iyyasu. For fuel, at first they used wood, especially *welgada* and *degeta* (*Carissa edulis*). These trees have mostly been cut, although there are still some of them around and they are regrowing. They also burned sorghum stalks; since the time of her grandfather, they have burned dung. There was no trouble with fuel when she was young, but there is now. While they still use dung for fuel, they use less of it than before. Nowadays, eucalyptus is the main fuel.

ASA was first married at Bur to HA from Abba Dam Ganda in Qatte, where they settled. They were married for about two years before they divorced at about the time the Italians left. ASA then married BM, also at Qatte. BM was *abagar* of that country and farmed *ma'im* (?) and *wafq* land, which is like *masgid* (mosque) land. At Qatte they planted *tef*, sorghum, beans, oats, chickpeas, vetch, barley, flax, lentils, peppers, and fenugreek. They also planted eucalyptus, *chat*, and coffee. ASA came to this particular location at the time of the Darg. BM had taken a second wife, and with land here, ASA left him. Only her youngest daughter was willing to come with her, while the other children stayed where they were. On her return, the Darg had started registering land. Since ASA with her daughter made a household of only two members, she lost land. The first redistribution left her with some of her father's land, but when they formed the producers' cooperative, that land was changed for the land of others. Neither ASA nor her daughter, who was too young, joined the cooperative. Having no cattle—"not even a hen to chase"—she let out her land on a fifty/fifty sharecropping basis and continues to do this.

Trees

There was no eucalyptus here before the Italians came; ASA first saw it at the compound of Fitawrari Bayyana, who was their Shum. They bought seedlings from him and used the wood for house construction, along with juniper and acacia. Originally they did not use eucalyptus for fuel, since there were many trees then. ASA only started to use eucalyptus as fuel after she came here. Here she has planted eucalyptus, euphorbia, a foreign type of *digita* (*Carissa edulis*), *welgada*, *qantafa* (*Pterolobium stellatum*, a leguminous shrub), *tasmania*, and foreign juniper. *Qantafa* she used for fencing; *tasmania* and *welgada* for house construction and fuel; juniper for house construction; the *digita* for fuel; and the eucalyptus for fuel, house construction, and to sell. As a member of the peasant association, she was able to get these seedlings free from the government nursery.

Famine

ASA coped with the help of a daughter at Gowana, who sent her money. She also received government relief—the government provided wheat, flour, oil, clothing, and butter, as well as money—and sold firewood at Bora. Lacking a husband, she was able to avoid resettlement, although some of her neighbors went, as did her brother's son, Ayyalaw, who was sent to Pawi, a lowland area. Her nephew thought that life was better there, where they had maize, rice, and government-supplied seeds. However, he was chased out of Pawi in 1986 Eth. Cal., after the coming of the new government, by the local people, the Shanqalla and the Shenasha, who massacred some 120 people. Now he has no land in spite of a letter from the zonal and the *warada* authorities explaining his plight. He is not a member of the peasant association and supports himself by working in the nursery and farming the land of others.

Family

ASA's father had five children, two boys and three girls, of whom the two boys died young. One of ASA's sisters lives at Qatte and the other at Zangetta, Abba Garoch Mandar. They are both married to farmers. ASA herself had ten children, five boys and five girls, of whom half died—two boys while young and three girls before marrying. One of her sons was drafted by the Darg and never returned. The other two are married at Qatte; one is a farmer and a *qallecha*, and the other is sick and cannot walk. Her two remaining daughters are also married: one at Gowane in Hararghe, the other to a teacher, who has taught at Asba Tafari, Awash, Meta Robi, and Waliso. This daughter now lives with ASA.

ASA and family

Life History of an Armachachaho (South Gondar) Farmer

AT

AT lives in the village of FA in Armachachaho, where he was born. He is sixty-seven years old. His father farmed *rest* land, not *qes maret*. AT's father had three oxen, three cows, one to five calves, eight goats, two donkeys, and one mule. There was no problem of grazing in the old days; people were few and animals could be driven anywhere below the mountain. When he was nineteen years old, in 1949 Eth. Cal., AT married Akalu Addalu Goshu from Janfanqara in Muna peasant association. They are still married.

Wild Animals

In the old days, wild animals—leopards, hyenas, and *affen*, a rare animal that kills by strangling—threatened domestic animals. These predators have all gone to the lowlands because the land has been cleared.

Fuel

In the old days he burned wood—*wanza* (*Cordia africana*), *tarme* (?), *qerkeha* (*Arundinaria alpina*, a kind of bamboo?), *bisanna* (*Croton macrostachys*), *qäðona*, *donga* (*Apodytes dimidiata*), *aqoma* (*Pygeum africanum*), *Shola* (*Ficus*), *kosheshele* (*Acanthus arboreus*). He still uses the wood of those trees and bushes. He used dung for kindling and sorghum stalks for torches, not as fuel. Today he has begun to burn sorghum stalks and feed them to the cattle. He burns more dung today.

Farming

AT stayed with his father for years after marrying, then set out farming on his own. He got a share of his father's land and some shares from others. "This area right here, was covered with forest (*dan*)," he said. AT cleared it and started to farm it. It was his *rest*. He paid a tithe to the government; he also paid one-fifth of his harvest to a *märét alaqa* (a landed authority, probably associated with a church). The actual payment varied from crop to crop, although the *alaqa* asked ten kilograms from each crop.

Crops

TA grew maize (*baqolo*), finger millet, *tef*, sorghum, and *nug* (*Guizotia abyssinica*). Only very recently have new varieties of crops, such as *tef* and sorghum, come. One example is "Canada" maize.

Trees and Bushes

TA grows *gesho* (*Rhamnus prinoides*), coffee, *chat*, bananas, and, most recently, papaya. He started to plant *gesho*, coffee, and banana when his father was still alive. He planted *chat* before the Darg came. He has also planted eucalyptus, but it does not do well because of the termites. TA also takes care of naturally occurring trees: "I pay good attention to them," he said. What was the forest like? In some places that area had been covered with trees, but not everywhere. It was mixed, with big trees, small trees, and bushes. In clearing the land, he cut everything down and burned it. A number of trees have been so reduced in numbers as to have almost disappeared, although none have actually done so. Eucalyptus was not new; Gondar was full of it. But there was no *qundo barbare* (*Mentha piperita*).

Government Tree Planting

There was no government-sponsored tree planting anywhere within sight. All the trees we see are protected by the peasant association. The *ballabat*s (including AT) overseers of land agreed to protect the trees.

Famine

From time to time problems prevailed, but the only famine here, according to some people, was in the Italian times.

Family

AT was one of twelve children—seven boys and five girls—one of whom, a boy, died young. Four of his five remaining brothers are dead now: one killed while searching for a lost ox; one, a priest, killed as part of a vendetta arising from the death of the other brother; one, a soldier, who went to Asmara; and one who was a farmer. Yet another brother was drafted by the Darg and returned here after sixteen years, with bullets in his body. AT's sisters all married, one of them as far away as Azazo, on the other side of Gondar.

AT himself had eleven children—five sons and six daughters. Four of them (two boys and two girls) were educated; one is too young to go to school, and one lives in Armachachaho-agade. Four of his daughters are married (one of them having been educated to sixth grade), and one is still a student.

Armachachaho landscape

Life History of an Armachachaho Farmer

EB

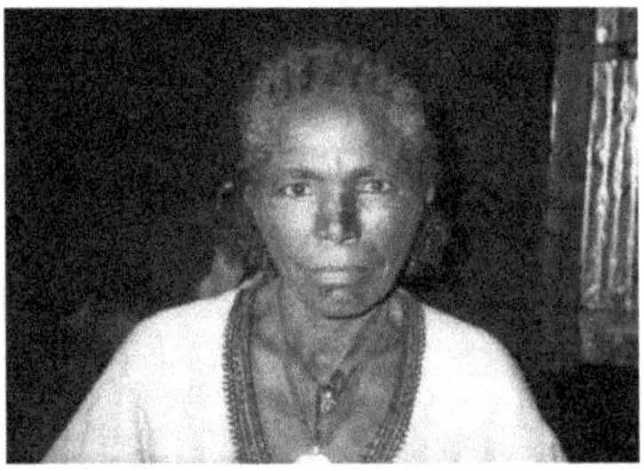

EB was born in the village of G in K peasant association in Armachachaho. The village is close to her current residence. Her father was a farmer on a tenure known as *laho*, which was the same as *rest*. He grew sorghum, *tef*, finger millet, barley, maize, and *nug*. He had four oxen, three to four cows with calves, two to three goats, and two donkeys. They had no problems finding grazing in those days on the *mask* or meadow. The number of people was very small. They helped each other. Today things are regulated and protected. Her father's household used indigenous wood for fuel; there was no eucalyptus then, but there were other trees: *dokma* (*Syzygium guineense*), olive, *aqakwora enchet* (*Buddleja polystachya*), *besana* (*Croton macrostachys*), *aqaqi* (*Maytenus ovatus*), *agam* (*Carissa edulis*), and *dediya*. They did not burn dung or crop residues because wood was plentiful, although they did collect some dung for baking *enjara*. Mostly they used wet dung to put on their soil, since it was good for growing corn and garden crops like *gesho*. EB's father had a few coffee bushes and some peaches, but there were no banana plants, lime trees, or any other fruit.

Marriage

EB married when she was twelve years old, around 1944 Eth. Cal. (1941–42), during Haile Selassie's time. She married in this village to DAT, who is still alive. He is a farmer, but under the Darg he joined the Ministry of Agriculture as a guard at its office. As long as he held the job, from which he has now retired, he lived in town, while she stayed here with the children. They farmed land she had inherited from her father as *rest*. Her husband was from elsewhere in the same peasant association. His father, A, worked in a government office and was a mounted soldier (*farasagna*). Others had taken his land.

EB and her husband had no oxen and no cattle, so DAT worked as *lole* (a servant of long standing), farming for others, getting one-quarter of the produce in return. He farmed for his uncle, his father's brother, AW. They begged oxen from people, and EB's father and brother helped them. They lived this way for thirty years. EB worked for others, weeding, while her husband farmed on a *wanfal* (cooperative) basis and got oxen from rich people in return. During the Darg they were able to get their own oxen but never received any kind of relief. To purchase the oxen, they sold donkeys and goats and sometimes grain, saved the money, and finally bought them. Now they have two oxen, one of which is weak, two cows, one calf, and one donkey. They graze these animals near the road. It is forbidden to graze beyond the road. Mostly they cut grass or weeds from their maize and sorghum fields and feed the cattle, which they keep tied up, just grazing them during the day. These days it is really a lot of trouble to find grazing, a "big thing."

Land Redistribution

This year (1997) the redistribution of land started in Magabit (March/April). Formerly she had had enough land to control her own grazing. In the previous redistribution she had received land according to the size of her family, some of which, the land she had used for grazing, she gave to her son. However, he became sick and went to Tekel Dengay. Gossipers told the peasant association committee about this and she lost all that land, "which was grabbed from us." She had been using some of it as a garden. She lost from three plots. The committee themselves took it and the secretary got some. She wept.

Trees

"There were no trees here when we started to farm. When we came here to farm there were only bushes—on the hillside beyond the road there were only bushes. There we also small seedlings. The trees you see now—*wanza* (*Cordia africana*), *besana*, *dedeya*, *dokma*, olive, eucalyptus—we planted." They started to grow eucalyptus under the Darg. Most of the trees they started by gathering seedlings from here and there. They use the trees as a windbreak, for fencing, for firewood, and in house construction. Eucalyptus and *wanza* are particularly useful for house construction, *dedeya* for fencing, and olive for house construction.

Bushes

"We planted *gesho*, but the soil is stony and unsuitable for it. We also planted coffee, but it died away," even though she nursed it for three years. They used to have peaches—and still do have some—but the birds eat the fruit. She also has a lemon-orange hybrid called *bakro*, which they eat. They planted bananas, but they did not grow.

Cash

Cash comes from selling goats, donkeys, field crops, and from her husband's job. They do not sell garden crops, although they do grow and eat them. They get no remittances. Her son, who is in town, cannot support himself—they support him.

Tree Planting

There was no tree planting during Haile Selassie's time. Under the Darg EB planted eucalyptus and received wheat in return—one *quintal* (100 kg) divided among eight people. There was a lot of wheat, but the officials kept most of it for themselves. She participated in planting all the eucalyptus, about a month each year for two or three months. Some of the planting was close to the road and today is protected by guards.

Famine

When she was a girl, she heard of a famine caused by hail that ruined their crops, but it took no lives. She also heard stories about the Kefu Qan, when there was a shortage of grains because of locusts and people died. Her mother's mother told her that many people died in this place. She did not hear of any other famines. She and her husband never experienced famine, although they did go through some hard times, which they coped with by buying grain and selling *gesho*. Some people had money set aside, some borrowed from others. One year, during the Darg, there was a famine because of army worms, which were connected to rain. *Tef* and finger millet were destroyed. The worm did not kill all the field crops, "we had a part harvest. We got pesticides from the Ministry of Agriculture in Tekel Dengay, but no relief."

Fuel

"Today we use forest wood—*dedeya, dokma*—whatever we can find." The use of dung has also increased.

Family

EB's mother was BK. Her father also had another wife, with whom he had three children. EB had four full siblings (two boys and two girls), of whom one of the girls died young. One of her brothers is a farmer on his father's land. He was forced to go to Humera for six months during the Green Revolution. Her other brother is in Gondar town, but they have lost touch with him. Two of her half brothers farm with their father and the third is a government employee in Gamu Gofa. EB and D had nine children in all (two boys and seven girls). One boy was a tailor, but he is now sick in town. The other is a third-grade student. Three of the girls were educated. One is in tenth grade; one went to ninth grade and then became a nun in Waldebba; and one finished twelfth grade, studied at the Teacher Training College at Qotabe, and is now a teacher in Gojjam. The other four girls all married farmers in this area.

Armachachaho landscape

Life History of a Dambeya Farmer

MKN

We interviewed MKN at the village of E in G peasant association in Dambeya. Dambeya is on the northeast shore of Lake Tana, belongs to the *wayna dagga* agro-climatic zone, and for almost 500 years has played an important role in providing agricultural products to the town of Gondar, capital of the Ethiopian kingdom in the seventeenth and eighteenth centuries. MKN was sixty-six years old at the time of the interview and had been born in a nearby peasant association.

MKN's father was a farmer, a *restennya* on *dabtara maret*, which meant that he paid rent to the church of Fit Mikael in Gondar. The rent was paid in kind, in grain—*tef*, sorghum, or finger millet. Her father was a rich man, with eight oxen, fifteen cows plus calves, thirty goats, three donkeys, and a mule. They grazed these animals on the communal meadow. There was plenty of land in those days, because farming was not so extensive. For fuel they used wood from trees—acacia, *wanza* (*Cordia abyssinica*), *dawda*, and *qemo*—and from bushes, like *Aqaqaa* (*Maytenus ovatus*) and *beshebsha* (*Cassia singueana*). Dung was not important as a fuel because of the amount of wood.

MKN first married during the Haile Selassie years, when she was seven years old, at Qobla. She stayed there with her mother- and father-in-law for about eight years, then lived with her husband for another couple of years, but they divorced because he was much older. She returned to her father's house, where she married BAZ; they are still married. They farmed together at Qobla for about twenty-seven years, before they came here sometime around 1974. As farmers they were *chisenya*, without inherited land, since neither she nor her husband had any. BAZ's father had had heritable land, but he died when BAZ was young. People gave them land, which they farmed on a three-to-one sharecropping basis. This was church land, *dabtara maret*, and the clerical owner got three parts of the harvest to their one part. A rich farmer provided them with the land, oxen, and seed. They grew *tef*, several kinds of sorghum, barley, wheat, and chickpeas.

For cash they sold field crops. They planted eucalyptus in their "backyard" (*gwaro*), land that BAZ's father gave them. They also planted *wanza* trees, as well as potatoes, onions, garlic, cabbage, sorghum, and maize—no coffee, *gesho*, *chat*, or fruit trees. They planted the trees for fuel, for house construction, and to sell. They sold to neighboring farmers, but not in town.

After they came here in 1966–67 Eth. Cal., at first they had no land. BAZ had a job at a local mill, and MKN sold beer (*talla*) and distilled alcohol (*araqey*). About five years later, under the Darg, in 1972 Eth. Cal. they received land from the peasant association. However, they had been better off as sharecroppers with only 25 percent of their harvest, than they were here, since their earlier land was more extensive and richer and yielded good returns. They grew the same crops here that they had grown in Qobla.

Livestock

For many years they had no oxen, although they did have three cows with calves, thirty sheep (which they sold in 1977 Eth. Cal.), and two donkeys, which they also sold. They graze their animals on common pastureland, but it is not enough. BAZ has to cut grass from field edges to feed the livestock. They still have two cows and two calves, and this year they bought an ox.

Cash

They primarily get cash from selling field crops, and they have sold animals. They used to have many sheep, but they sold them all, because they had no children to look after them, and they got a good price. BAZ worked at the mill for seventeen years and now is employed as a guardian at the school. MKN continues to sell *talla* and *araqey*, and people come from the neighborhood to buy it. (Some came during the interview and MKN insisted we have some of her *talla*.) MKN used to distill the *araqey* herself, but now she brings it from Gondar. She buys the supplies for making *talla*—sorghum, maize, and *gesho* in nearby Qolla Duba town—and brings it here by car.

Tree Planting

"We planted eucalyptus. I haven't counted them, but they are many." They also planted *wanza*, *gesho*, and coffee, as an experiment (*lamayat*, to "see"). MKN did not plant enough *gesho* for her talla business, because she is "lazy." They have to fetch the water to make both *talla* and coffee. They have grown *gulo* (also called *bulka*—*Ricinus communis*—the castor oil plant), which they sell to tanners. They have planted euphorbia.

Famine

There has been no real famine, but there have been shortages. During Italian times, in 1931 Eth. Cal., there was drought and maybe locusts. Every four years there is a recurrent problem of flooding and then drought. Sometimes the water is polluted, and then malaria becomes a serious problem.

Family

MKN has ten half siblings—seven boys and three girls. Of them, five boys and one girl died young. MKN's two surviving half brothers are married and farmers in Dambeya. Of the half sisters who reached maturity, two are now dead [numbers per original field notes]. The surviving sister never married. She studied to ninth grade and now lives by selling *talla*. MKN and BAZ had just one girl, who is now a policewoman in Dabra Tabor.

Plowing in Dambeya

Life History of a Dambeya Farmer

MW

MW still lives in the G, the peasant association where he was born, sixty-eight years before we interviewed him.

Childhood, Marriage, and Family

MW was married in *Genbot* 1939 Eth. Cal. (May/June 1946 AD) to KEG from Sufanqara Kidana Mehrat, which is in Dambeya. At the time KEG was about seven years old. When we pressed him on the precision of this date, he went into his house and brought out his handwritten journal, which he had maintained for much of his life. MW and KEG are still married. First they lived in his father's house for about seven years.

MW's mother was Abbäbä Mäšäša. MW was one of seven siblings, all boys. Two, who had been farmers, are now dead. Two others live and farm here, as their brothers had. One brother is a teacher in Dessie and one teaches in Qolla Deba.

MW had ten children, six boys and four girls. Two of his sons—one of them a teacher and the other a statistician—died from tuberculosis. One son is in third grade here, and three others are farmers, of whom two can write and the third reached sixth grade. One daughter is a teacher at Esté; one still lives at home, having completed twelfth grade; and two, not educated, are married to farmers.

MW has never lived elsewhere, although he has gone to Addis Ababa on official business. In 1963 Eth. Cal. he made a trip sponsored by the Ministry of Agriculture to Assala in Arsi to see about animal farming and seedlings, improved seeds, and fertilizer use.

MW taught himself to read and write Amharic. In 1945 Eth. Cal. he had been very sick and was taken to a healing spring at Mwat in Alafa. He had had an experience with a friend who had bought an ox. The sale was later contested, and to defend his case, the friend went to a priest to have a letter prepared. This convinced MW of the value of reading and writing.

One year after the Italians came, there was a famine caused by locusts. There was also one under Haile Selassie caused by drought. Traders brought sorghum from Humera, he sold his livestock and bought the grain. One year during the Darg, maybe in 1977 Eth. Cal., there was severe drought and famine. Even last year there was drought. Here they are not affected by flooding.

Farming

His father farmed *rest* land. They had no *dabtara* land, but they did farm *rim* land, which supported the Mass at the Church of Fit Mikael, Gondar. Their land was part of the foundation lands of the church, and MW and his father paid one-half *madaga* to the church. They also paid *asrat* to the government, which amounted to seven birr a year in Haile Selassie's time but rose to twenty under the Darg. They generated cash by selling their field crops in the market. MW's father grew *tef*, chickpeas, sorghum and some finger millet, and some barley.

MS's father had four or five oxen, three or four cows plus calves, ten goats, two donkeys, and a mule. For grazing during the rainy season, they would drive the cattle to Deba Mountain in the high country above Gondar town. During the rest of the year they grazed on extensive public meadowland near Lake Tana. For some time now, people have settled on the meadowland and are growing sorghum, potatoes, maize, and so on. The area tends to flood during the rains, but it drains during the dry season, and people moved in.

MW started farming for himself before he married. After he married, his father gave him some land to farm. He also farmed empty land near his home. He started to pay taxes about this time. He had two oxen, which he and his wife received as a marriage endowment, one from his father and one from her father, and six cows, three from each side of the family. Later MW raised sheep. He grew the same crops as his father. He has two kinds of soil—*walka* and *qäy* (red). Under Haile Selassie and the Darg he was introduced to three or four new kinds of *tef,* all of which came ultimately from the national agricultural research center at Dabra Zayt, near Addis Ababa. Farmers were able to pick those which most suitable to them. The Darg also introduced a new kind of sorghum, but it has not been successful here.

MW owns two oxen, one cow, one calf, and one donkey. His mule has been sold.

Land redistribution started in 1972 Eth. Cal. and MW lost land. Forced villagization, under the direction of government teams who came from Gondar and Qolla Deba, took place in 1979. "At that time we moved here, near to my farm." He lost his earlier garden but planted a new one here. Others who had not lost their gardens returned to their old home sites when the Darg collapsed.

MW has used fertilizer ever since his trip to Arsi in 1963. It was given freely to use with improved seeds during Haile Selassie's days. Many people adopted its use following MW's example. About fifty-three foreigners and Ethiopians honored him by coming to observe his use of fertilizer. However, today it is very expensive and "we might have to abandon it. Last year I sold an ox to repay my loan from a previous purchase of fertilizer and this year I won't use it." Many people sold oxen to repay their fertilizer loans, including his son and his brother.

To generate cash, MW sold field crops and, later, eucalyptus.

Malaria affects many people. It used to be very serious. Under Haile Selassie they sprayed DDT. After Haile Selassie, the government stopped spraying. During the Darg, local people made numerous petitions for renewed spraying, but it only happened a few times. Malaria is increasing again, and people are dying from it.

As far as trees go, in the old days there were *warka* (*Ficus vasta*), *wanza* (*Cordia africana*), *šola* (*Ficus sur*), *dokma* (*Szygium guineense*), *senat* (*Myrica salicifolia*), *šenšena* (?), acacia, *besanna* (*Croton macrostachys*), *agam* (*Carissa edulis*), *qaga* (*Rosa abyssinica*), and *enkoyi* (*Ximenia americana*)—many different kinds of trees, all indigenous and very thick. During Italian times MW was a herder and often had difficulty finding his way while driving the livestock because of the density of the vegetation. Today the trees are almost totally gone. *Dokma*, *qaga*, *agam* have disappeared. On Deba Mountain you can find some *besanna* and other trees. No new trees have come, although some foreign vegetables have come.

Fuel

In those days they used the wood of traditional (*dero*) trees: acacia, *besanna*, and *wanza*; also bushes like *agam*. They used very little dung and only during the rains. For fuel today, the use of dung has increased. They prepare it during the dry season and use it mostly on the *me*ᵃ*ad* for baking *enjära*. Wood, especially from eucalyptus trees, is important for fuel. They cut branches and dry them and chop down the trees and prepare logs.

Tree Planting

MW's maternal uncle first planted eucalyptus around here, in 1930 Eth. Cal., during the Italian occupation. He brought seedlings from Guramba and demonstrated them to people. MW knows the first people to grow eucalyptus, and they did so before Italian times. It is not clear whether they brought the seedlings from Gondar or Dabra Tabor, or even Addis Ababa.

Around 1953 Eth. Cal., during Haile Selassie's time, MW planted eucalyptus, year by year adding to his plantation; under the Darg, he added *Ficus thoningii* and *wanza*. MW also planted *gesho* but lost it without compensation during land redistribution in 1972 EC. He has never planted coffee, *chat*, or fruit trees.

"We started planting trees under Haile Selassie, along the edge of the road. I, myself, participated in government tree planting under both regimes," he explained. The soil is not good, where the authorities wanted trees planted, "and we told them not to plant," but they went ahead anyway, and those trees dried up. The Darg also encouraged them to plant trees for themselves and distributed seedlings: "That's how we got these eucalyptus trees." They also planted a protected state forest at Aege, but at the fall of the Darg, people cut it down.

NOTES

Introduction

1. Pauline Boerma, "Seeing the Wood for the Trees: Deforestation in the Central Highlands of Eritrea since 1890" (PhD diss., Oxford University, 1999), and her more recent article based on her early 1990s fieldwork and the Italian photographic archives, "Myths, Memories, and Metaphors: Recollecting Landscape Changes in the Eritrean Highlands," *Journal of Eastern African Studies* 6, no. 2 (2012): 246–69.

2. See Kebede Tadesse, *Trees of Ethiopia: A Photographic Guide and Description* (Addis Ababa: Washera, 2012). This text, in Amharic and English, is a comprehensive compendium of trees found in Ethiopia by agroecological zone, local names, and scientific classification.

Chapter 1. Famine in Ethiopia: The Great Famine

1. A group of five informants interviewed with the assistance of Merere Ejjeta in Tekel Dengay, Lay Armachachaho, on August 6, 1997. The average age of the group was fifty-nine years. One of the informants had a son who spent time doing relief work in Wallo in 1984. (For more information on the informants, please see the appendix.) McCann's informants also remembered the Great Famine of 1888: James McCann, *From Poverty to Famine in Northeast Ethiopia: A Rural History, 1900–1935* (Philadelphia: University of Pennsylvania Press, 1987), 30; as did an informant of Andrew J. Carlson and Dennis G. Carlson, *Health, Wealth, and Family in Rural Ethiopia: Kossoye, North Gondar Region, 1963–2007* (Addis Ababa: Addis Ababa University Press, 2008), 96.

2. Richard Pankhurst remains the only scholar to give this event extended consideration. His now classical account is "The Great Ethiopian Famine of 1888–1892: A New Assessment," *Journal of the History of Medicine and Allied Sciences* 21 (1966): 95–124 and 271–94. Pankhurst updated the story in *The History of Famine and Epidemics in Ethiopia Prior to the Twentieth Century* (Addis Ababa: Relief and Rehabilitation Commission, 1986); and in Richard Pankhurst and Douglas H. Johnson, "The Great Drought and Famine of 1888–92 in Northeast Africa," in *The Ecology of Survival: Case*

Studies from Northeast African History, ed. David M. Anderson and Douglas H. Johnson (Boulder, CO: Westview Press, 1988), 47–70. Pankhurst's studies, with the extraordinary reach of their primary source citations, have been the point of departure for my exploration of the Kefu Qan, which draws on additional primary material and a relevant secondary literature, which has yet to be brought fully into the discussion. For some of the additional primary sources, see particularly the works edited by Bairu Tafla: *A Chronicle of Emperor Yohannes IV (1872–89)* (Wiesbaden: Franz Steiner, 1977), *Asma Giyorgis and His Work: History of the Galla and the Kingdom of Šawa* (Stuttgart: Frank Steiner, 1987), and *Ethiopian Records of the Menilek Era: Selected Amharic Documents from the Nachlass of Alfred Ilg 1884–1900* (Wiesbaden: Harrassowitz, 2000). Several European sources not used by Pankhurst have been edited by Carlo Zaghi: Augusto Salimbeni, *Crispi e Menelich nel Diario inedito del conte Augusto Salimbeni: Con Prefazione, Introduzione, Note e Appendici* (Turin: Istituto per la storia del Risorgimento italiano, 1956); Augusto Salimbeni, *La Conquista dell'Africa: Studi e Ricerche*, 2 vols. (Napoli: Istituto Universitario Orientale, 1984); and J. Voellmy, *Arthur Rimbaud: Correspondance 1888–1891* (Paris: Gallimard, 1965). For relevant secondary material, see Richard Caulk, *"Between the Jaws of Hyenas": A Diplomatic History of Ethiopia 1876–1896*, ed. Bahru Zewde (Wiesbaden: Harrassowitz, 2002); John Ford, *The Role of Trypanosomiases in African Ecology: A Study of the Tsetse Fly Problem* (Oxford: Clarendon Press, 1971); and C. A. Spinage, *Cattle Plague: A History* (New York: Kluwer Academic/Plenum, 2003).

3. For accounts of the complicated political situation, see Sven Rubenson, *The Survival of Ethiopian Independence* (London: Heinemann Educational Books, 1976); Zewde Gabre-Sellassie, *Yohannes IV of Ethiopia: A Political Biography* (Oxford: Clarendon Press, 1975); Caulk, *"Between the Jaws"*; and Haggai Erlich, *Ethiopia and Eritrea during the Scramble for Africa: A Political Biography of Ras Alula, 1875–1897* (East Lansing: African Studies Center, Michigan State University, 1982). For issues of chronology, a valuable tool is Chris Prouty Rosenfeld, *A Chronology of Emperor Menilek II of Ethiopia* (East Lansing: African Studies Center, Michigan State University, 1976).

4. Ford, *Trypanosomiases*, 139. There had been recorded outbreaks of rinderpest in the lower Nile valley. The principal symptoms of rinderpest are fever, runny nose, and acute diarrhea; Spinage, *Cattle Plague*, 3–5

5. The rinderpest raged southward from the Ethiopian highlands, triggering disaster in Maasailand and great hardship all the way to South Africa and westward through Sudanic Africa. See, for example, John Lonsdale, "The European Scramble and Conquest in African History," *The Cambridge History of Africa*, vol. 6, *1870–1905*, ed. Roland Oliver and G. N. Sanderson (Cambridge: Cambridge University Press, 1985), 689–90; and Ford, *Trypanosomiases*. In 1897 it reached Ovamboland in northern Namibia; E. Kreike, *Re-creating Eden: Land Use, Environment and Society in Southern Angola and Northern Namibia* (Portsmouth, NH: Heinemann, 2004), 37–41.

6. Let Marefia, January 14, 1892, *Bolletino della Società Geografica Italiana*, S. 3, vol. 5 (1892): 231. This is a good deal more plausible than Salimbeni's suggestion that "the empire's population had been reduced to a third since 1888–89," for which see Caulk, *"Between the Jaws,"* 357. The estimate of one-third human mortality is seconded by the French medical expert R. Wurtz, who was sent in 1897 by the French government

to investigate the rinderpest epidemic; R. Wurtz, "Hygiène Publique et Privée en Abyssinie," *La Semaine Médicale* (1898): 489–94.

7. Harvey Marcovitch, *Black's Veterinary Dictionary*, 41st ed. (London: A&C Black, 2005). Contemporary sources on Ethiopia are consistent with such numbers.

8. Mangestu Lamma, *Mashafa Tezeta ZaAlaqa Lamma Haylu Walda Tarik* (Addis Ababa, 1959 Eth. Cal.), 142.

9. *Bolletino*, S. 3, vol. 5 (1892): 227.

10. For example, *Bolletino della Società Geografica Italiana.*

11. For example, Bairu Tafla, *Ethiopian Records of the Menilek Era.*

12. For example, for the Catholics, see *Les Missions Catholiques*; for the Protestants, Johann Martin Flad, *60 Jahre in der Mission unter den Falaschas in Abessinien: Selbstbiographie des Missionars Johann Martin Flad: Mit Einleitung und Schlusswort von seinem Sohn Pastor Friedrich Flad* (Giessen and Basel, 1922).

13. For example, PRO, Foreign Office, FO 403, passim.

14. For Harar, see PRO, FO 403; and Voellmy, *Arthur Rimbaud.*

15. For Gabra Sellasie, see *Chronique du règne du Ménélik II, Roi des Rois d'Éthiopie, d'après un Manuscrit Original*, ed. and trans. Casimir Mondon-Vidailhet, 2 vols. (Paris, 1930, 1932); for Asma Giyorgis, see Bairu Tafla, *Asma Giyorgis and His Work.*

16. *Les Missions Catholiques* 21 (1889): 62.

17. Caulk, *"Between the Jaws,"* 104–5.

18. Antonelli to the interim Minister of Foreign Affairs, Ankober, May 10, 1888, Camera dei Deputi, Documenti presentati al Parlamento Italiano, Seduta del 17 Dicembre 1889 (Roma, 1890), 326. *Caleb* is the Amharic *qalab*, "nourishment, food." I am indebted to Irma Taddia of Bologna University for making these materials available to me.

19. Antonelli to the interim Minister of Foreign Affairs, Let-Marefia, June 10, 1888, Camera dei Deputati, 336.

20. Antonelli, Aussa, December 10, 1888, Camera dei Deputati, Documenti Diplomatici, Seduta del Dicembre 1889, Etiopia (Roma, 1890), 383; Rosenfeld, *Chronology*, 133.

21. See, for example, Antonelli to the interim Minister of Foreign Affairs, Hadelò Gubo (Aussa), Decembre 9, 1888, Camera dei Deputati, 379. Neither Caulk nor Zewde, mention hunger in the army at this time.

22. *Les Missions Catholiques* 20 (1888): 557. An identical extract from the letter was also published in *Annales de la Propagation de la Foi* 61 (1889): 143–44.

23. Gustav Arén, *Evangelical Pioneers in Ethiopia: Origins of the Evangelical Church Mekane Yesus* (Stockholm: EFS Förlaget; 1978), 319. Arén's account rests on missionary sources: the archives of the Swedish Evangelical Mission together with its periodical, *Missions Tidning*; and on two retrospective accounts, E. Winqvist, *Med livet som insats: Läkaremissionen i Eritrea* (Stockholm, 1945), and K. Winqvist, "En liten återblick på min 25-åriga verkshamet som läkaremissionär," *Varde Ljus* (1912): 38–61.

24. Pankhurst, "Ethiopian Famine," 112, citing a letter in *Missions Catholiques* of 1889. A year later, on January 20, 1890, Crouzet wrote from Massawa, "It hasn't been very long, a month at most, since we have seen people dying of famine and misery." *Missions Catholiques* 22 (1890): 86.

25. "Tous les fléaux," *Missions Catholiques* 21 (1889): 62.

26. Baldissera to the Minister of War, Italian Chamber of Deputies, Documenti presentati al Parliamento Italiano, Etiopia, "L'occupazione di Keren e dell'Asmara," Sedute del 17 dicembre 1889 (Roma, 1890), 41, 43. Rosenfeld asserts that as of May 1889, Massawa was full of hungry refugees; *Chronology*, 137.

27. Count Antonelli to the Interim Minister of Foreign Affairs, Adis Abeba, February 20, 1889, Italian Chamber of Deputies, Documenti Diplomatici presentati al Parliamento Italiano, Etiopia, Sedute del 17 dicembre 1889 (Rome, 1890), 399–400.

28. Caulk, *"Between the Jaws,"* 202.

29. Mangestu, *Mashafa Tezeta*, 106.

30. "Micidiale epidemia tifica," Ragazzi to the Geographical Society, Lat Marafeya, August 14, 1889; *Bolletino*, S. 3, vol. 2 (1889): 964–65.

31. Rimbaud to Ilg, Harar, August 26, 1889; Voellmy, *Rimbaud: Correspondance*, 103.

32. Rubenson, *Survival*, 384; Zewde, *Yohannes IV*, 249.

33. Gabra Sellasie, *Chronique du règne du Ménélik II*, 2:296. See note 49 for the view of another Ethiopian author, Asma Giyorgis.

34. Caulk's translation (*"Between the Jaws,"* 203) of Ilg to Rimbaud, Antotto, October 26, 1889; Voellmy, *Rimbaud: Correspondance*, 138.

35. Abba Fezzah to Abba Elias, Kafa, April 1892; *Missions Catholiques* 41 (1909): 303. Abba Fezzah wrote, "A great calamity having broken out in 1889 in Kafa, destroying the cattle and causing the deaths of many people, I preached to everyone, during those evil days, the necessity of receiving baptism, and of making penance in order to appease the divine wrath."

36. *Les Missions Catholiques* 22 (1890): 85–86. Salimbeni reported that there was also smallpox in Massawa at this time. Augusto Salimbeni, *Crispi e Menelich nel Diario inedito del conte Augusto Salimbeni: Con Prefazione, Introduzione, Note e Appendici*, ed. Carlo Zaghi (Turin: Istituto per la storia del Risorgimento italiano, 1956); Augusto Salimbeni, *La Conquista dell'Africa: Studi e Ricerche*, ed. Carlo Zaghi, 2 vols. (Napoli: Istituto Universitario Orientale, 1984), 20.

37. *Les Missions Catholiques* 22 (1890): 350.

38. Leopoldo Traversi, *Let-Marefià: Prima Stazione geografica Italiana nelo Scioa e le nostre Relazioni con l'Etiopia (1876–1896)* (Milan, 1931), 333. This is the same epidemic mentioned by de Lauribar, who identifies it as "cholera" and has it running through July 1890. Paul de Lauribar, *Douze Ans en Abyssinie (Souvenirs d'un Officier)* (Paris, 1898), 193. Wurtz identifies it as "influenza": "Hygiène Publique et Privée en Abyssinie," 492.

39. Traversi, *Let-Marefià*, 355.

40. Bairu Tafla, *Ethiopian Records of the Menilek Era*, 393.

41. Salimbeni, *Crispi e Menelich nel Diario*, 71.

42. Ibid., 96.

43. Until October 1895: Rosenfelt, *Chronology*, 141–73. See also Gabra Sellasie, *Chronique du règne du Ménélik II*, 2:264–65.

44. Bairu Tafla, *A Chronicle of Emperor Yohannes IV*, 190–91.

45. Salimbeni, *Crispi e Menelich nel Diario*, 129.

46. Flad, *60 Jahre*, 385.

47. Caulk, *"Between the Jaws,"* 237; citing a letter in the Italian archives from Salimbeni, Entotto, November 14, 1890. See also Traversi, *Let-Marefià*, 335. A number of Ethiopian sources mention cannibalism: Wudu Kassu, "References to the Environment in Nineteenth Century Ethiopian Chronicles," unpublished paper, referring to Alaqa Lamma, to the Gojjami chronicler Alaqa Takla Iyaus, and to Afa Warq Gabra Iyasus.

48. Bairu Tafla, *Ethiopian Records of the Menilek Era*, 399.

49. Asma also placed famine in the preceding year of Mark. Bairu Tafla, *Asma Giyorgis and His Work*, 828–29.

50. Voellmy, *Rimbaud: Correspondance*, 207–8.

51. Salimbeni, *Crispi e Menelich nel Diario*, 313, 319.

52. Ferdinando Martini, Lettere [1860–1928] con 26 Tavole e 7 Autografi fuor di Testo (Milan, 1934), 249.

53. Conrad Keller, *Alfred Ilg: Sein Leben and sein Wirken als schweizerischer Kulturbote in Abessinien* (Frauenfeld and Leipzig: Huber, 1918), 89–90.

54. PRO, FO 403/155, 47, Consul Stace to Brigadier-General Jopp, Aden, June 30, 1891. The Russian traveler Mashcov passed through Dardar at about this time and observed that a country, which on his previous visit three years ago had been rich and fertile, "has been converted into an uninhabited desert." O. Cerroti, "Il secondo Viaggio in Abissinia del Mashcov (1891–1892)," *Bolletino della Società GeograficaItaliana*, S. III, vol. 7 (1894): 850.

55. *Bolletino della Società Geografica Italiana*, S. 3, vol. 4 (1891): 1009.

56. PRO, FO 403/155, 65 Consul Stace to Sir E. Baring, Aden, October 31, 1891, reporting on a conversation two days earlier with a Mr. McDonald.

57. PRO, FO 403/147, 1–2, Stace to Dufferin, Rome, January 6, 1892.

58. Caulk, *"Between the Jaws,"* 238.

59. Caulk, *"Between the Jaws,"* 296, citing British reports from Aden.

60. Keller, *Alfred Ilg*, 89–90; Bairu Tafla, *Ethiopian Records of the Menilek Era*, 413, Yoséf Negusé to Ilg, Entoto, September 14, 1891. Another correspondent reported similarly with respect to Shawa, a few months later: Wälättä Maryam to Ilg, Entoto, April 12, 1892, in Bairu Tafla, *Ethiopian Records of the Menilek Era*, 420. For Crouzet's letter, see the material referenced in note 24.

61. Leopoldo Traversi, Lat Marafeya, January 14, 1892, *Bolletino della Società Geografica Italiana*, series 3, vol. 5 (1892): 229–30. For the Oromo lands, see the retrospective account by Russian traveler Alexander Bulatovich, *Ethiopia through Russian Eyes: Country in Transition 1896–1898*, trans. and ed. Richard Seltzer (Lawrenceville, NJ: Red Sea Press, 2000). Traversi's reference to Harar is supported by the "Report by Lieutenant-Colonel Stace on the Trade and Commerce of the Somali Coast Protectorate for the Year 1891–92," 7, 9–10: Foreign Office, 1893, Annual Series, No. 1208, Diplomatic and Consular Reports on Trade and Finance, Africa, Report for the year 1891–92 on the Trade of the Somali Coast Protectorate, 1893. The references to Harar are also supported retrospectively by Sylvain Vignéras, *Une Mission française en Abyssinie* (Paris, 1897), 56, 83.

62. Bairu Tafla, *Ethiopian Records of the Menilek Era*, 422, Gabra Sellase Neguse to Ilg, Entoto, April 25. For the Amharic text, see 130.

63. Letter of Fr. Picard, Akrur, July 22, 1892; *Les Missions Catholiques* 24 (1892): 438.

64. Georg Schweinfurth, *Einige Mitteilungen über seinen Diesjährigen Besuch in der Colonia Eritrea (Nord-Abessinien): Vortrag Gehalten in der Gesellschaft für Erdkunde zu Berlin Am 2 Juli 1892* (Berlin, 1892), 4. For Crouzet, see *Annales de la Propagation de la Foi* (1892): 31. There are two retrospective references to smallpox: Vignéras, *Mission française*, 80; and Wurtz, "Hygiène Publique et Privée en Abyssinie," 492. Wurtz was a French medical researcher, heavily dependent on information from Alfred Ilg.

65. G. Schweinfurth, *Einige Mitteilungen*; Consul Barnham to Sir E. Baring, Suakin, April 25, 1892, Document 20, 137–39, in *British Documents on Foreign Affairs: Reports and Papers from the Foreign Office Confidential Print*, part I, *From the Mid-Nineteenth Century to the First World War, Series G, Africa, 1885–1914*, vol. 13, *Abyssinia and Its Neighbours, 1854–1914* (Frederick, MD: University Publications of America, 1995). For the specific reference, see 138. See also Picard's letter of July 22, 1892, *Les Missions Catholiques* 24 (1892), 438.

66. Flad, *60 Jahre*, 410–11, 413. Flad quotes from two different letters here, the latter dated August 24, 1892 (i.e., toward the end of the rainy season), the former presumably a month or two earlier. "Unbeschreiblich" is the term repeated in Flad's accounts, once in reference to 1890.

67. FO 403/177, 16, Consul Barnham to Baring, Suakin, April 25, 1892.

68. FO 403/177, 13, Stace to Bering, Aden, April 12, 1892. This expedition is also recorded by Gabra Sellasie, *Chronique du règne du Ménélik II*, 322.

69. Carlo Zaghi, *La Conquista dell'Africa: Studi e Ricerche*, 2 vols. (Napoli: Istituto Universitario Orientale, 1984), vol. 2, 812, Capucci to Salimbeni, Ankobar, September 5, 1892.

70. Ibid., 817, October 5, 1892, "il paese è affamatissimo."

71. Ibid., 818, from Addis Abeba, December 27.

72. J. Theodore Bent, *The Sacred City of the Ethiopians Being a Record of Travel and Research in Abyssinia in 1893* (London, 1896), 87, 91. Other cholera references are retrospective: de Lauribar, with reference to Eritrea from late 1889 into the first half of 1890, *Douze Ans*, 193; Vignéras, writing of Harar in 1890, *Mission française*, 56; and, finally, Wurtz, "Hygiène Publique et Privée," 492.

73. Bairu Tafla, *Asma Giyorgis and His Work*, 805.

74. Pankhurst's various publications give a clear account of the information provided by these travelers.

75. Lonsdale, "European Scramble and Conquest," 689.

76. For effects elsewhere in Africa, see, in addition to Lonsdale, Shula Marks, "Southern and Central Africa, 1886–1910," in *Cambridge History of Africa*, vol. 6, *From 1870 to 1905*, ed. Roland Oliver and G. N. Sanderson (Cambridge: Cambridge University Press, 1985), 436, 446–47, 467; Marcia Wright, "East Africa, 1870–1905," in *Cambridge History of Africa*, vol. 6, *From 1870 to 1905*, ed. Roland Oliver and G. N. Sanderson (Cambridge: Cambridge University Press, 1985), 572; Richard Waller, "Emutai: Crisis and Response in Maasailand 1883–1902," in *Ecology of Survival: Case Studies from Northeast African History*, ed. Douglas Johnson and David Anderson (London: I.B. Tauris, 1991), 73–112; and Ford, *Trypanosomiases*, passim.

77. See the accounts of Bettembourg and Arén. Shula Marks, writing of southern and central Africa, tells of rinderpest preceded by smallpox and followed by locusts, drought, and other epidemic human diseases. Marcia Wright, in her account of the same period in eastern Africa, notes how the turmoil surrounding colonial subjugation was compounded by "locusts, rinderpest, drought, smallpox, sleeping sickness and other diseases and pests." Marks, "Southern and Central Africa, 1886–1905," 425, and Wright, "East Africa, 1870–1905," 572.

78. For all issues relating to Ethiopian agronomy, see H. P. Huffnagel, *Agriculture in Ethiopia* (Rome: FAO, 1961); and E. Westphal and J. M. C. Westphal-Stevels, *Agricultural Systems in Ethiopia* (Wageningen: Centre for Agricultural Publishing and Documentation, 1975).

79. This is not lowland as might commonly be understood. The lower limit of *qolla* is somewhere around 500 meters above sea level. Below it lies *baraha*, desert or wilderness.

80. An excellent source of relevant information is the *Atlas of the Ethiopian Rural Economy* published in 2006 in Addis Ababa and Washington, DC, by the Central Statistical Agency in Addis Ababa in association with the Ethiopian Development Research Association and the International Food Policy Research Institute. For "traditional" agroclimatic zones, see 15.

81. On June 28, 2011, the Food and Agricultural Organization of the United Nations formally declared that the rinderpest virus had been eradicated, only the second disease to be so designated and the first animal disease. "Rinderpest Eradicated—What Next?," http://www.fao.org/news/story/en/item/80894/code. See also "Rinderpest, Scourge of Cattle, Is Vanquished," *New York Times*, June 28, 2011.

82. *Oxford English Dictionary*, 2nd ed. (Oxford: Oxford University Press, 1989). The earliest source of which I am aware to use the term "rinderpest" in connection with Ethiopia is Augustus B. Wylde, *Modern Abyssinia* (London, 1901), 339.

83. See Karl Appuhn, "Ecologies of Beef: Eighteenth-Century Epizootics and the Environmental History of Early Modern Europe," *Environmental History* 15, no. 2 (2010): 268–87, for an account of the Venetian republic's response to the spread of rinderpest within the context of new markets for and sources of beef in Western Europe.

84. Marcovitch, *Black's Veterinary Dictionary*.

85. Spinage, *Cattle Plague*, 504, argues that this was not the first episode in Ethiopia and the disease was not introduced by the Italians in 1887 but earlier, possibly by the Egyptians in 1876. Ford (*Trypanosomiases*, 138, 394–95) is equivocally inclined to the opposite view, that the Italians were the agents and Massawa the point of entry in 1887.

86. For Yohannes and Sahai, see above.

87. Pankhurst draws effectively for the early chronology of the spread of the epidemic on the memory of Alaqa Lamma, most particularly as gleaned in interview by Lamma's son, Mangestu. Pankhurst, "Great Ethiopian Famine," 102. Lamma was blessed with a remarkable memory. In my grasp of Alaqa Lamma's information, I am indebted not only to Pankhurst but also to Wudu Kassu, "References to the Environment." See also Alaqa Lamma, "Tarik Yasme Takla Giyorgis[enna] Yatse Yohannes," Mss. Éthiopien d'Abbadie 259, folio 36 (Bibliothèque Nationale, Paris, n.d.), which confirms Tigray by June. The informants of Abdussamad H. Ahmad confirm Lamma's

information, although not precisely because they allow for an earlier appearance of the disease in Gojjam. They claimed, "By the beginning of 1889 most of the cattle in Gojjam had died." Abdussamad H. Ahmad, "Peasant Conditions in Gojjam during the Great Famine 1888–1921," *Journal of Ethiopian Studies* 20 (November 1987): 1–18. The July–December hiatus, as evident from their table I.3, is puzzling. To be sure, the rains greatly inhibited all kinds of movement from late June to early September, which accounts for these months. But how to account for the September–December hiatus is unclear.

88. See the account by Antonelli. Antonelli's claim that "entire herds of 500 and 1000 head have been destroyed in one or two days" echoes Alaqa Lamma's recollection.

89. Jack R. Harlan, *The Living Fields: Our Agricultural Heritage* (Cambridge: Cambridge University Press, 1995), 217–18. See also Huffnagel, *Agriculture in Ethiopia*; Westphal and Westphal-Stevels, *Agricultural Systems in Ethiopia*; Frederick J. Simoons, *Northwest Ethiopia: Peoples and Economy* (Madison: University of Wisconsin Press, 1960), chap. 5 (64–84); A. Hoben, *Land Tenure among the Amhara of Ethiopia: The Dynamics of Cognatic Descent* (Chicago: University of Chicago Press, 1973), especially 54–58; and Dan Franz Bauer, *Household and Society in Ethiopia* (East Lansing: African Studies Center, Michigan State University, 1977), chap. 3 (57–77). Hoben and Bauer offer anthropologically informed local views from Dagga Damot in Gojjam, Endarta, and Tigray, respectively.

90. Hans Ruthenberg, *Farming Systems in the Tropics*, 2nd ed. (Oxford: Clarendon Press, 1976), 135, table 6.3. The data in the table date from 1969 to 1973.

91. André G. Haudricourt and Mariel Jean-Brunhes Delamarre, *L'Homme et la charrue à travers le monde* (Paris, 1955), offers an account of the evolution of the plow and its changing physical characteristics from its adoption into modern times, with some comments on how the Ethiopian plow (122–27, 301–2) fits into this picture. E. Hahn, *Die Entstehung der Pflugkultur (unsres Ackerbaus)* (Heidelberg, 1909), offers a cultural history.

92. The flavor of the debate may be gathered from the following sources: Christopher L. Delgado and John McIntire, "Constraints on Oxen Cultivation in the Sahel," *American Journal of Agricultural Economics* 64, no. 2 (1982): 188–96; Eric W. Crawford and Gregory C. Lassiter, "Constraints on Oxen Cultivation in the Sahel: Comment," *American Journal of Agricultural Economics* 67, no. 3 (1985): 684–85; Christopher L. Delgado and John McIntire, "Constraints on Oxen Cultivation in the Sahel: Reply," *American Journal of Agricultural Economics* 67, no. 3 (1985): 686–87; Kimseyinga Savadogo, Thomas Reardon, and Kyosti Pietola, "Farm Productivity in Burkina Faso: Effects of Animal Traction and Nonfarm Income," *American Journal of Agricultural Economics* 76 (1994): 608–12; and Timothy O. Williams, "Problems and Prospects in the Utilization of Animal Traction in Semi-Arid West Africa: Evidence from Niger," *Soil and Tillage Research* 42, no. 4 (1997): 295–311. I am indebted to Alex Winter-Nelson for directing me to this material. Theoretical issues are addressed by Prabhu Pingali, Yves Bigot, and Hans P. Binswanger, *Agricultural Mechnization and the Evolution of Farming Systems in Sub-Saharan Africa* (Baltimore: Johns Hopkins University Press for the World Bank, 1987).

93. For the presence of rinderpest in Shawa as of February, see Antonelli's letter of February 20, 1889. In the north, the epidemic struck sometime in the months

following March 1888, possibly before the onset of intensive field preparation when Yohannes marched south with diseased cattle in his train. Notices of ensuing famine are November in Eritrea and February for provinces from Tigray southward.

94. Gabra Sellasie, *Tarik zaman za Dagmawi Menilek negusa nagast za-Ityopya* (Addis Ababa, 1959), 146 (146 in the Amharic, 264 in the French translation). See also Gabra Sellasie, *Chronique du règne du Ménélik II*, chap. 40. For humans yoked to the plow, see Pankhurst, "Ethiopian Famine," 120, on the basis of oral information from Alaqa Lamma. For attempted adaptations in Gojjam, see Ahmad, "Peasant Conditions," 5. Ahmad notes that horses were customarily used to draw plows in Agawmeder, a district within Gojjam, where horses were plentiful.

95. Gabra Sellasie, *Tarik*, 175 (296 in the translation); for the emperor's setting an example of hoe cultivation, see 177 (297). See also Salimbeni, *Crispi e Menelich nel Diario*, 129; and Afawarq Gabra Iyasus, *Dagmawi Ase Menilek* (Rome: n.p., 1927), 64.

96. Online sources include www.britannica.com and Wikipedia. The desert locust is the object of concern of the intergovernmental organization Desert Locust Control Organization for Eastern Africa, based in Addis Ababa and established by an international convention signed in 1962. Member countries are Djibouti, Eritrea, Ethiopia, Kenya, Somalia, Sudan, Tanzania, and Uganda.

97. Several sources describe this combination, most interesting being Menilek's proclamation of July 1889, for which see the account by Ragazzi, note 30.

98. Bairu Tafla, *Ethiopian Records of the Menilek Era*, 413.

99. Crouzet, *Les Missions Catholiques* 21 (1889), 62; and Arén, *Evangelical Pioneers in Ethiopia*, 319.

100. I am inclined not to consider "dysentery" a distinct disease but a condition, which might have been produced by a variety of factors. The evidence for typhus is elusive.

101. *Les Missions Catholiques* 22 (1890), 99; Crouzet, Massawah, January 20, 1889; Pankhurst, "Ethiopian Famine," 121; Bairu Tafla, *Asma Giyorgis and His Work*, 803; Gabra Sellasie, *Tarik*, 177; Capucci to Salimbeni, Entotto, October 5, 1892; Salimbeni, *Conquista dell'Africa*, 817.

102. "La typhus de famine vint encore ajouter ses ravages à ceux que le choléra et la variole avaient déjà faits," *La Semaine Médicale*, 492. In this case, "typhus" is explicitly distinguished from cholera. For Stace, see PRO, FO 403/155, 65, Consul Stace to Sir E. Baring, Aden, October 31, 1891.

103. The Amharic sources are of little help here. Both Gabra Sellasie and Asma Giyorgis speak generally of *bashata*, undifferentiated epidemic disease. Asma Giyorgis mentions *taqmat*, or diarrhea (Bairu Tafla, *Asma Giyorgis and His Work*, 804–5); and Gabra Sellasie speaks of *ma'at*, "horror; wrath of God; . . . calamity," according to Amsalu Aklilu, *Amaregna-Englizegna Mazgaba Qalat/Amharic-English Dictionary* (Addis Ababa: Kuraz Asātāmi Drjt, 1979 EC), 25. Amsalu renders *basheta* "disease, illness." Amsalu, *Amaregna-Englizegna*, 96. Ilg's correspondents alone of the Amharic sources contemporary with the Kefu Qan use a specific term, *koléra*: Bairu Tafla, *Ethiopian Records of the Menilek Era*, letters 59, 93, 95, and 96.

104. Richard Pankhurst, "The Hedar Bäšeta of 1918," *Journal of Ethiopian Studies* 13, no. 2 (1975): 103–31 (103 for the case in point); Richard Pankhurst, "The History

and Traditional Treatment of Smallpox in Ethiopia," *Medical History* 9 (1965): 343–55. For the dictionary, see I. Guidi, *Supplemento al Vocabolario Amarico-Italiano Compilata con il Concorso di Francesco Gallina ed Enrico Cerulli* (Rome: Istituto per l'Oriente, 1940), col. 6. Nor was this a problem unique to Ethiopia. The entry on "influenza" in Wikipedia claims that the symptoms of the great 1918 pandemic were "so unusual that initially influenza was misdiagnosed as dengue, cholera or typhoid."; See http://en.wikipedia.org/wiki/Influenza#History (accessed December 2, 2010). Pankhurst makes the general point that there is great difficulty in identifying diseases mentioned in Ethiopian sources prior to the nineteenth century and the advent of trained medical observers, although he is not notably strict in realizing the implications of the point. Richard Pankhurst, *An Introduction to the Medical History of Ethiopia* (Trenton, NJ: Red Sea Press, 1990). See also, in addition to the previously cited articles: Richard Pankhurst, "The History of Cholera in Ethiopia," *Medical History* 12 (1968): 262–69; Richard Pankhurst, "Some Notes for the History of Typhus in Ethiopia, *Medical History* 20 (1976): 384–93; and Richard Pankhurst, "A Historical Note on Influenza in Ethiopia," *Medical History* 21 (1977): 195–200.

105. See C. W. Potter, "A History of Influenza," *Journal of Applied Microbiology* 91, no. 4 (2001): 572–79, for an account of the great pandemics; also *Encyclopaedia Britannica*, 11th ed. (1902–11), vol. 6, 262–67, for cholera; and vol. 16, 552–56, for influenza. Pankhurst makes a plausible case that both smallpox and influenza were endemic.

106. Our source for the edict is Ragazzi's report of August 14, 1889, from Lat Marafeya as published in the *Bolletino*. Bairu Tafla, *Asma Giyorgis and His Work*, 804–5. Asma goes on to say, "Both the people and the animals also became insane. The cattle were extinct; but not too many people died."

107. Menilek's expedition is well documented. See, for example, Gabra, *Tarik*, chaps. 43 and 44; Caulk, *"Between the Jaws,"* 200ff.; Rosenfeld, *Chronology*, 140–43.

108. Traversi, *Let-Marefià*, 333.

109. Caulk, *"Between the Jaws,"* 237, n.14, citing from the Italian archives the dispatch by Count Salimbeni to Crispi, Entotto, October 8, 1890.

110. For 1890, see Salimbeni, *Crispi e Menelich nel Diario*, 20; for January 1892, Schweinfurth, *Einige Mitteilungen*, 4; and for July 1892, Crouzet from Akalä Guzay, July 2, 1892, *Missions Catholiques* 24 (1892): 438. For Wurtz on the endemic status of smallpox, see "Hygiène Publique et Privée," 492.

111. PRO, FO 403/155, 65 Consul Stace to Sir E. Baring, Aden, October 31, 1891; Traversi to Società Geografica Italiana, *Bolletino*, S. 3, vol. 4 (1891): 1009: "Carestia, di giorno in giorno più acuta, e tifo fanno stragi."

112. For Ilg's correspondents, see Bairu Tafla, *Ethiopian Records of the Menilek Era*, 421, Makwannen Walda Mika'el to Ilg, Deder, April 16, and 422, Gabra Sellase Neguse to Ilg, Entoto, April 25. See also Caulk's citation of a report of Salimbeni also for April, which claimed that cholera in Harar had run its course: Caulk, *"Between the Jaws,"* 326.

113. Flad, *60 Jahre*, 413; Capucci to Salimbeni, Ankobar, September 5, 1892, in Salimbeni, *Conquista dell'Africa*, 812. Bent, traveling in Eritrea at the beginning of 1893, found plenty of evidence of earlier cholera epidemics. *Sacred City*, 11, 87, and 91.

114. Pankhurst, "The Hedar Bäšeta of 1918."

115. R. A. Caulk, "Armies as Predators: Soldiers and Peasants in Ethiopia c. 1850–1935," *International Journal of African Historical Studies* 11, no. 3 (1978): 457–93.

116. In addition to the sources cited already, see also Bahru Zewde, *A History of Modern Ethiopia 1855–1991*, 2nd ed. (Oxford: James Currey, 2001), 42–55.

117. For the importance of Boru Meda in the reconstruction of the state, see Donald Crummey, "Imperial Legitimacy and the Creation of a Neo-Solomonic Ideology in 19th Century Ethiopia," *Cahiers d'Études africaines* 109, no. 28 (1988): 13–43; and Donald Crummey, "Church and Nation: The Ethiopian Orthodox *Täwahedo* Church (from the Thirteenth to the Twentieth Century)," in *Cambridge History of Christianity*, vol. 5, *Eastern Christianity*, ed. Michael Angold (Cambridge: Cambridge University Press, 2006), 457–87.

118. The best account is Erlich, *Ethiopia and Eritrea*. But see also Zewde, *Yohannes IV*. A primary source for local power struggles in Hamasén is Johannes Kolmodin, *Traditions de Tsazzega et Hazzega*, 2 vols. (Upsala, Appelberg, 1915–16).

119. For an account of some of these developments, see Caulk, *"Between the Jaws,"* 89–93.

120. See Richard Caulk, "Religion and the State in Nineteenth Century Ethiopia," *Journal of Ethiopian Studies* 10, no. 1 (1972): 23–41, who emphasizes the violence of Wallo in the mid to late 1880s. But for a different account, see Zewde, *Yohannes IV*, 94–100.

121. Bahru, *History of Modern Ethiopia*, 59; Zewde, *Yohannes IV*, 246–47; Caulk, *"Between the Jaws,"* 106–26.

122. The Egyptians had also occupied Harar, and in 1887, following their withdrawal, Menilek incorporated it into the empire. Rimbaud perceived the occupation as harsh (Voellmy, *Rimbaud: Correspondance*, 102–3, 107) and contributing to the suffering of the town with the onset of the Kefu Qan (letters from Harar of August 26, 1889, and September 7, 1889).

123. Gabra Sellasie, *Tarik*, chap. 39. Gabra Sellasie gives no clear date for this edict, but it clearly preceded Menilek's second edict.

124. Gabra Sellasie somewhat confusingly sets this action following the rains of 1889. *Tarik*, chap. 46.

125. Ragazzi, Lat Marafeya, August 14, 1889; *Bolletino*, S. 3, vol. 2 (1889): 964–65.

126. Ragazzi, Entotto, November 16, 1889; Camera dei Deputati, Documenti Diplomatici, Etiopia (Serie 2a), Seduta del 6 maggio 1890 (Roma, 1890), 52. Marcus claims that large feasts were held, "notable for their economy, out of consideration for the famine conditions in much of the country" (Harold G. Marcus, The *Life and Times of Menelik II: Ethiopia, 1844–1913* (Lawrenceville, NJ: Red Sea Press, 1995), 117. For Gabra Sellasie's account, see *Tarik*, chap. 40.

127. For people seeking relief, see Gabra Sellasie, *Tarik*, chap. 39; also chap. 46. Taytu was Menilek's wife. There are particularly vivid and gruesome accounts of people from Tigray flocking to Eritrea, and Eritrean highlanders flocking to Massawa: for April–May 1890, see Salimbeni, *Crispi e Menelich nel Diario*, 70–71; see also Traversi, *Let-Marefià*, 355; Traversi on Menilek and Taytu, letter to Società Geografica, Lat Marafeya, January 14, 1892, *Bollettino*, S. 3, vol. 5 (1892): 231.

128. Wurtz, "Hygiène Publique et Privée," 492. Unfortunately, none of these caravans succeeded, having been plundered en route by the Somali and Afar, who were equally famine-stricken.

129. Pankhurst, "Ethiopian Famine," 284.

130. Gabra Sellasie, *Tarik*, chap. 46.

131. Pankhurst, "Ethiopian Famine," 286; Wudu, "References to the Environment."

132. Gabra Sellasie, *Tarik*, chap. 52. I am indebted to Wudu Tafete's reading of this material. But see also the account in Donald Crummey, *Land and Society in the Christian Kingdom of Ethiopia: From the Thirteenth to the Twentieth Century* (Urbana: University of Illinois Press, 2000), 222. For a brief summary of the discussion and reference to the principal sources, see Crummey, *Land and Society*, chap. 10, n. 111.

133. PRO, FO 403/147, pp. 1–2, Stace to Dufferin, Rome, January 6, 1892.

134. Caulk, *"Between the Jaws,"* 238. Arsi, an Oromo area southeast of Addis Ababa, was one such area.

135. The classical, still valuable account is Sven Rubenson, "Adwa, 1896: The Resounding Protest," in *Protest and Power in Black Africa*, ed. Robert Rotberg and Ali Mazrui (New York: Oxford University Press, 1970), 113–42. See also Rubenson, *Survival*, 399–406; and Abdussamad H. Ahmad and Richard Pankhurst, eds., *Adwa Victory Centenary Conference, 26 February–2 March 1996* (Addis Ababa: Institute of Ethiopian Studies, Addis Ababa University, 1998).

136. Ferdinando Martini, *Nell'Affrica Italiana* (Milan: Fratelli Treves, 1925), 25. This view was decidedly not shared by Salimbeni, who, on April 30, 1890, writing of the Eritrean plateau between Asmara and Däbarwa, claimed that "now nobody is cultivating, because nobody believes that the Italian occupation here will be something stable." Salimbeni, *Crispi e Menelich nel Diario*, 70. Salimbeni, who saw nothing but desolation up to this point, judged that Saqota, in Wag, was reasonably prosperous at this same time; Salimbeni, *Crispi e Menelich nel Diario*, 99.

137. See, for example, Traversi's letter of December 23, 1891, dated Lat Marafeya; *Bollettino*, S. 3, vol. 5 (1892): 225–26.

138. Cerroti, "Il secondo Viaggio," 850; Voellmy, *Rimbaud: Correspondance*, 102–3.

139. Crouzet, writing from Keren at Easter 1891, *Missions Catholiques* 23 (1891): 393; Picard, Akrur, July 24, 1892, *Missions Catholiques* 24 (1892): 438.

140. Menilek was probably himself Ethiopia's leading merchant. Caulk, *"Between the Jaws,"* chaps. 1 and 2.

Chapter 2. Famine in Ethiopia, 1973–1974 and 1984–1985

1. Lillian M. Li, "Famine and Famine Relief: Viewing Africa in the 1980s from China in the 1920s," chap 18 in *Drought and Hunger in Africa: Denying Famine a Future*, ed. Michael Glantz (Cambridge: Cambridge University Press, 1987), 415–34.

2. "Victim" never gets defined in the literature. A working definition, fitting the case at hand, is a person made desperate by a collapse in the availability of food. There were many more victims than there were deaths from famine. Many of the famine deaths came from contagious diseases and from exposure in the relief camps.

3. Mesfin Wolde Mariam, *Rural Vulnerability to Famine in Ethiopia 1958–1977* (New Delhi: Vikas Publishing, 1984).

4. The famine of the Kefu Qan had fairly precise dates—outbreak following the rinderpest pandemic of mid-1888 and ending with a satisfactory harvest in 1892. In the famine of 1973–74 in northeast Ethiopia, the peak of starvation occurred in 1973, and the peak in Hararghe was 1974. The peak of starvation in the 1980s occurred primarily in 1984 and 1985. In Amharic the famine of the 1980s is known as the famine of Saba-sabat (the year 1977 in the Ethiopian calendar, which runs from September 1984 to September 1985), indicative of the peak of its intensity.

5. Amartya Sen, *Poverty and Famines: An Essay on Entitlement and Deprivation* (Oxford: Clarendon Press, 1981), 86. De Waal, citing the same sources, put the number of dead between 40,000 and 60,000; Alex de Waal, *Famine Crimes: Politics and the Disaster Relief Industry in Africa* (London, IN: African Rights and the International African Institute, 1997), 106. Given that the famine revealed how poorly information was generated and transmitted in Ethiopia in the early 1970s, estimates of its scale necessarily varied widely and are unlikely ever be more firmly grounded.

6. For the nearest precedents to 1973, see Mesfin, *Rural Vulnerability*, v and 35–36. The larger story of famine in Ethiopian history has not been carefully researched, but there are accounts by Bahru Zewde, "A Historical Outline of Famine in Ethiopia," in *Rehab: Drought and Famine in Ethiopia*, ed. Abdul Meijid Hussein (London: International African Institute, 1976), 52–58; Pankhurst, *The History of Famine and Epidemics*; and Workineh Degefu, "Some Aspects of Meteorological Drought in Ethiopia," in *Drought and Famine in Africa: Denying Famine a Future*, ed. Michael Glantz (Cambridge: Cambridge University Press, 1987), 29–31, table 2.1. For a detailed study of famine in Tigray in the third quarter of the twentieth century, see Adhana Haile Adhana, "History of Selected Famines in Peasant Societies in Tigray and Wallo, Ethiopia, 1941–1974" (PhD diss., Addis Ababa University, 1996).

7. Mesfin's account (*Rural Vulnerability*, chap. 2) draws on the findings of a national Commission of Inquiry, which he chaired, into the origins of the famine of 1972–74. For the quotation, see 49. For the commission, whose report was never published, see 40 and note 61.

8. Abdul Mejid, "The Political Economy of the Famine in Ethiopia," in *Rehab: Drought and Famine in Ethiopia*, ed. Abdul Meijid Hussein (London: International African Institute, 1976), 23.

9. J. P. W. Rivers, J. F. J. Holt, J. A. Seaman, and M. R. Bowden, "Lessons for Epidemiology from the Ethiopian Famines," *Annales de la Société belge de médecine tropicale* 58, nos. 4–5 (1976): 350–51. For summaries of the crop survey's findings, see table 3 on 353; see also Sen, *Poverty and Famines*, 92–95; and Mejid, "The Political Economy of the Famine in Ethiopia," 23–32.

10. Mesfin, *Rural Vulnerability*, 49. See also the account by G. Kumar, *Ethiopian Famines 1973–1985: A Case-Study* (Helsinki: World Institute for Development Economics Research, 1987), chap. 3.

11. Jack Shepherd, *The Politics of Starvation* (New York: Carnegie Endowment for International Peace, 1975), 15 and 20.

12. Mesfin documents in some detail the government's refusal to even accept, let alone act on, information about conditions in Wallo. Mesfin, *Rural Vulnerability*, 44–51. See also Rivers et al., "Lessons for Epidemiology," 352.

13. Rivers et al., "Lessons for Epidemiology," 351.

14. Ibid., 351–52; Julius Holt and John Seaman, "The Scope of the Drought," in *Rehab: Drought and Famine in Ethiopia*, ed. Abdul Meijid Hussein (London: International African Institute, 1976), 4; Shepherd, *Politics*, 22. See also Kumar, *Ethiopian Famines*, 11.

15. For a vivid account of conditions in Wallo and Tigray in 1974, see Liam Nolan, *The Forgotten Famine* (Dublin: Mercier Press, 1974).

16. *New York Times*, November 18, 1973.

17. Shepherd, *Politics of Starvation*, 39. See also Rivers et al., "Lessons for Epidemiology," 351–52. There is little information about the famine in Bagemder and Gamu Gofa. Bagemder is immediately to the west of Wallo, and affected districts probably were those along the Wallo border. Gamu Gofa is in southern Ethiopia on the highlands west of the Rift Valley.

18. The nature of Haile Selassie's government is well captured by Christopher Clapham, *Haile-Selassie's Government* (London: Longmans, Green, 1969); John Markakis, *Ethiopia: Anatomy of a Traditional Polity* (Oxford: Clarendon Press, 1974); and Patrick Gilkes, *The Dying Lion: Feudalism and Modernization in Ethiopia* (New York: St. Martin's Press, 1975). For accounts of the man himself, see Peter Schwab, *Haile Selassie I: Ethiopia's Lion of Judah* (Chicago: Nelson Hall, 1979); Harold G. Marcus, *Haile Sellassie I: The Formative Years, 1892–1936* (Berkeley: University of California Press, 1987); and Hans Wilhelm Lockot, *The Mission: The Life, Reign and Character of Haile Sellassie I* (London: Hurst, 1989).

19. Gilkes, *Dying Lion*, 76–78. For the revenue changes, see Crummey, *Land and Society*, 237–41.

20. Markakis, *Anatomy*, 150.

21. For Menilek's introduction of the Council of Ministers, see Bahru Zewde, *A History of Modern Ethiopia*, 114–16; for Haile Selassie's government, see Bahru Zewde, *A History of Modern Ethiopia*, 201–9.

22. "An Oriana Fallaci Interview: Journey into the Private Universe of Haile Selassie," *Chicago Tribune*, June 24, 1973, A1. The interview was conducted in French, the European language Haile Selassie knew best. I was directed to this source by L. Wiseberg, "An International Perspective on the African Famines," in *The Politics of Natural Disaster: The Case of the Sahel Drought*, ed. M. H. Glantz (New York: Praeger, 1976), 108. For the reference to Shepherd, see *Politics*, 42. Mesfin also has an account of the visits to Wallo; see *Rural Vulnerability*, 42–43.

23. There is a voluminous literature on the outbreak and unfolding of the revolution. Three early accounts were Colin Legum, *Ethiopia: The Fall of Haile Selassie's Empire* (New York: Africana, 1975); Marina Ottaway and David Ottaway, *Ethiopia: Empire in Revolution* (New York: Africana, 1978); and René Lefort, *Ethiopia, an Heretical Revolution?* (London: Zed, 1983). One of the best of a later wave of books is Andargachew Tiruneh, *The Ethiopian Rrevolution 1974–1987: A Transformation from an Aristocratic to a Totalitarian Autocracy* (Cambridge: Cambridge University Press, 1993). For other valuable accounts, see John Markakis and Nega Ayele, *Class and Revolution in Ethiopia* (Nottingham: Spokesman, 1978); Fred Halliday and Maxine Molyneux, *The Ethiopian Revolution* (London: Verso, 1981); John Harbeson, *The Ethiopian Transformation: The*

Quest for the Post-imperial State (Boulder, CO: Westview Press, 1987); Christopher Clapham, *Transformation and Continuity in Revolutionary Ethiopia* (Cambridge: Cambridge University Press, 1988); and Edmond Keller, *Revolutionary Ethiopia: From Empire to People's Republic* (Bloomington: Indiana University Press, 1990).

24. Ottaway and Ottaway, *Empire in Revolution*, 3; Lefort, *Heretical Revolution*, chap. 2.

25. Andargachew provides a particularly clear view in *Ethiopian Revolution*, chap. 2.

26. Ottaway and Ottaway, *Empire in Revolution*, 2.

27. Lefort, *Heretical Revolution*, 48; Gilkes, *Dying Lion*, 169.

28. Sen, *Poverty and Famines*, 92–95. For additional summaries, see Mejid, "Political Economy"; and Rivers et al. "Lessons for Epidemiology." Mejid's summary is the most detailed. The three sources agree that at the national level the harvest of 1972–73 was adequate to cover drought-induced shortfalls in Shawa, Wallo, and Tigray.

29. Holt and Seaman made their definitive case in Julius Holt and John Seaman, "Markets and Famines in the Third World," *Disasters* 4, no. 3 (1980): 283–97; with earlier iterations in Holt and Seaman, "Scope of the Drought"; and in Rivers et al., "Lessons for Epidemiology," 355. Sen appears to have been mistaken about prices in the Dessie market, which he presented as having been only modestly affected. Den, *Poverty and Famines*, 92–95. Peter Cutler refined the argument in "Famine Forecasting: Prices and Peasant Behavior in Northern Ethiopia," *Disasters* 6, no. 1 (1984): 48–56. In a study of drought and impending famine in Ethiopia in 1989–90, Derseh Endale concluded that the relationship between crop failure and the price of grain in local and regional markets was ambiguous, clearly inflationary for some types of grain and without impact in the case of others. The underlying question is: does crop failure increase effective demand, or decrease it because of its effect on farmers' incomes? Derseh Endale, *Rural Markets, Food-Grain Prices and Famines: A Study on Selected Regions in Ethiopia*, Working Paper no. 118 (Washington, DC: United Nations University World Institute for Development Economics Research, 1995).

30. Ottaway and Ottaway, *Empire in Revolution*, 56–57; and Lefort, *Heretical Revolution*, 68–69.

31. Shepherd, *Politics of Starvation*, 50–52.

32. Gilkes, *Dying Lion*, 248–51; Randi R. Balsvik, *Haile Sellassie's Students: The Intellectual and Social Background to Revolution, 1952–1977* (East Lansing: African Studies Center, Michigan State University, 1985).

33. Clapham, *Transformation and Continuity*, 49.

34. Bahru, *History of Modern Ethiopia*, 195.

35. For the English text of the land nationalization proclamation, see the appendix to "Rural Development in Ethiopia," a special issue of *Rural Africana* 28 (Fall 1975); and the appendix to Kidane Mengisteab, *Ethiopia: Failure of Land Reform and Agricultural Crisis* (New York: Greenwood Press, 1990). For assessments of the proclamation, see John W. Bruce, "Ethiopia: Nationalization of Rural Lands Proclamation, 1975," *Land Tenure Center Newsletter* 47 (January–March 1975): 1–15; Dessalegn Rahmato, "Agrarian Reform in Ethiopia: A Brief Assessment," in *Proceedings of the Seventh International Conference of Ethiopian Studies*, ed. Sven Rubenson (Addis Ababa: Institute of Ethiopian Studies, 1984), 587–95; Dessalegn Rahmato, *Agrarian Reform in Ethiopia*

(Trenton, NJ: Red Sea Press, 1985), chap. 3; and Andargachew, *Ethiopian Revolution*, 97–112.

36. There are many accounts. See particularly Lefort, *Heretical Revolution*, chap. 3; and Andargachew, *Ethiopian Revolution*, 98–101 and 157. The Ottaways published the text of the Peasant Association Consolidation proclamation as appendix B (217–27) of *Empire in Revolution*.

37. Dessalegn, *Agrarian Reform*, chap. 4.

38. Ibid., 66–68; Clapham, *Transformation and Continuity*, chap. 7, gives a first-rate succinct account of Darg agricultural policy. For the AMC, see particularly 168–71; for the military draft, 109–10.

39. For the political movements in opposition to the Darg, see John Young, *Peasant Revolution in Ethiopia: The Tigray People's Liberation Front, 1975–1991* (Cambridge: Cambridge University Press, 1997); for Eritrea, see especially Ruth Iyob, *The Eritrean Struggle for Independence: Dominance, Resistance, Nationalism 1941–1993* (Cambridge: Cambridge University Press, 1995). For the wars of the Darg, see Gebru Tareke, *The Ethiopian Revolution: War in the Horn of Africa* (New Haven, CT: Yale University Press, 2009).

40. Thomas J. Bassett and Donald Crummey, "Contested Images, Contested Realities: Environment and Society in African Savannas," in *African Savannas: Global Narratives and Local Knowledge of Environmental Change*, ed. Thomas J. Bassett and Donald Crummey (Oxford: James Currey, 2003), 1–30.

41. Lester R. Brown and Edward C. Wolfe, *Reversing Africa's Decline*, Worldwatch paper 65 (Washington, DC: Worldwatch Institute, 1985), 3, an expanded and republished version of their 1978 paper. Since 1984 the Worldwatch Institute has annually published a State of the World book series.

42. Yeraswork Admassie, *Twenty Years to Nowhere: Property Rights, Land Management and Conservation in Ethiopia* (Lawrenceville, NJ: Red Sea Press, 2000), xxii.

43. Alemneh Dejene is a senior officer in the UN FAO Sustainable Development Department and author of *Environment, Famine, and Politics in Ethiopia* (Boulder, CO: Lynne Rienner, 1990). For the claim about soil conservation, see Alemneh Dejene, *Integrated Natural Resources Management to Enhance Food Security: The Case for Community-Based Approaches in Ethiopia* (Rome: UN FAO, 2003), 6. Dessalegn Rahmato has nuanced the discussion of official thinking about the environment in "Environmentalism and Conservation in Wallo before the Revolution," *Journal of Ethiopian Studies* 31, no. 1 (1998): 43–86, which establishes the heroic figure of Dajzmach Mammo Seyum, chief administrator of Wallo from 1964 to 1970, who anticipated almost the entire package of environmental interventions, which was developed following 1974.

44. Hurni is director of the Swiss Centre for Environment and Development and editor of *Mountain Research and Development*.

45. As noted already, precise dates are problematic. I follow the popular Ethiopian designation of the year as Saba-sabat, 1977 Eth. Cal. (1984–85 AD).

46. "Famine Confronts 150 Million in Africa," *New York Times*, March 29, 1981. For a general chronology of the run-up to the famine of 1984, see Kurt Jansson, Michael Harris, and Angela Penrose, *The Ethiopian Famine* (London: Zed Books, 1987), 133–39.

47. Peter Gill, *A Year in the Death of Africa: Politics, Bureaucracy and the Famine* (London: Paladin, 1986), 17–18.

48. "Drought Pushes More Ethiopians into Relief Camps," *Christian Science Monitor*, March 14, 1983. This story was filed from Ebinat in Gondar province. The US embassy in Addis Ababa on the basis of satellite imagery projected a 2-million-ton deficit. The Ethiopian government agency issuing the appeal was presumably the RRC, whose commissioner, Shimelis Adugna, is mentioned later in the *Monitor* article.

49. "Food Aid to Ethiopia Said to Be Misused," *Washington Post*, March 28, 1983; "Europeans to Continue Aid to Ethiopia Despite Charges," *Washington Post*, March 19, 1983.

50. "Surge of Ethiopian Refugees Threatens the Horn of Africa," *Christian Science Monitor*, April 26, 1983; see also "Ethiopia, in Grips of Severe Drought, Has 'Critical' Need of Food, Supplies," *Baltimore Sun*, May 2, 1983. For a study of Ethiopian refugees in the Sudan, see Gaim Kibreab, *People on the Edge in the Horn* (Oxford: James Currey, 1996).

51. "Wanted: Trucks to Carry Food to Drought-stricken Ethiopians," *Christian Science Monitor*, July 28, 1983; "US Pledges $3 Million in Food Aid to Ethiopia," *Washington Post*, July 29, 1983.

52. "Food Aid Stymied as Ethiopia Wars Drag On," *Christian Science Monitor*, October. 25, 1983.

53. *The Sunday Times*, December 4, 1983. Kurt Jansson, who became head of the UN Relief Operation in Ethiopia, angrily denied these charges. Jansson, Harris, and Penrose, *Ethiopian Famine*, 68–73. For another charge, see "Ethiopians Accused of Aid Charade," *The Times* (London), January 24, 1984. de Waal, *Famine Crimes* (123–27), repeats these charges under the headings "The Big Lie" (123) and "Militarized Relief" (124).

54. Associated Press, "Ethiopian Drought Drags On and On," *Champaign-Urbana News-Gazette*, February 26, 1984. The story bore the subtitle "7.3 million faced food shortage."

55. Dawit Wolde Giorgis, *Red Tears: War Famine and Revolution in Ethiopia* (Trenton, NJ: Red Sea Press, 1989), 135, 151; Jansson, Harris, and Penrose, *Ethiopian Famine*, 139–53. For a press report, see *The Times* (London), May 4, 1984, which reported that "the lives of 5.2 million Ethiopians, most of them in Eritrea, Tegray, and Wallo," were at stake. Ten days later, *The Times* reported that "between five and seven million" Ethiopian lives were at risk, from "the worst drought in Ethiopia's history." Dawit Wolde Giorgis had served the Darg as "a senior member of the Marxist-Leninist Party," deputy foreign minister, chief representative in Eritrea as of 1979 (Dawit, *Red Tears*, 1, 93).

56. *The Guardian*, July 13, 1984; *Baltimore Sun*, October 25, 1984.

57. For a journalist's eye-witness account, see Gill, *A Year in the Death*, 7–10; see also Dawit, *Red Tears*, 170–76; Bahru, *Modern History*, 255–56; and Clapham, *Transformation and Continuity*, 190–91.

58. Clapham, *Transformation and Continuity*, 191–92; Jansson, Harris, and Penrose, *Ethiopian Famine*, 153–54. The British broadcasts were on October 23 (BBC) and October 25 (Thames Television).

59. Jansson, Harris, and Penrose, *Ethiopian Famine*, 74; Dessalegn Rahmato, *Famine and Survival Strategies: A Case Study from Northeast Ethiopia* (Uppsala: Nordiska

Afrikainstitutet, 1991), 108–9. Bahru has 1 million: *Modern History*, 263; Dawit says 1.2 million, *Red Tears*, 355. Patrick Webb and Joachim von Braun, *Famine and Food Security in Ethiopia: Lessons for Africa* (Chichester: John Wiley & Sons, 1994), 18, say: "Excess mortality between 1983 and 1986 in Ethiopia and Sudan probably exceeded 1 million." At its height, in 1985, relief was channeled through 286 distribution centers, 162 feeding stations, and 3 major shelters; Edmond J. Keller, "Drought, War and the Politics of Famine in Ethiopia and Eritrea," *Journal of Modern African Studies*, vol. 30, no. 4 (1992), 617.

60. Dessalegn, *Famine and Survival*, 204–5, reports many deaths from cholera in Wallo in the first half of 1985.

61. Dessalegn, *Famine and Survival*, 101.

62. Gill, *A Year in the Death*, 30–31 persuasively argues for the reliability of RRC figures, specifically those that accompanied the major appeal of March 1984. de Waal, *Famine Crimes*, 112–15, offers the most trenchant criticism of their reliability. No numbers have supplanted those provided by the RRC, which were accepted by Kurt Jansson and his colleagues and are at least useful for general scale and a rough comparison province to province. For continuing citation, see Fassil G. Kiros, *Enough with Famines in Ethiopia: A Clarion Call* (Addis Ababa, 2005). Fassil accepts the numbers and cites the Disaster Preparedness and Prevention Commission, set up by the government that overthrew the Darg in 1991, as doing so as well; see *Enough with Famines*, table 3.1, 78.

63. Ethiopia, Office of the Population and Housing Census Commission, *1984 Population and Housing Census Preliminary Report*, vol. 1, part 1 (Addis Ababa, 1984). Of particular relevance to this study are provincial reports: *Analytical Report on Gondar Region* (Addis Ababa, 1990), and *Analytical Report on Wello Region* (Addis Ababa, 1991). What bearing the unfolding famine had on the conduct of the census is an untold story. For the results of the census, see also Clapham, *Continuity and Transformation*, 129. A sample survey conducted by the Central Statistical Office in 1970 estimated total population to be about 33.7 million.

64. Dessalegn notes that the value of food aid was self-reported by the countries involved and likely to have been somewhat inflated. However, even reducing the figure by 10 or 20 percent still makes it sizable compared with the total expenditure of the central government of Ethiopia at this time. Göte Hansson, *The Ethiopian Economy 1974–94: Ethiopia Tikdem and After* (New York: Routledge, 1995), 35, 115, and 167.

65. Gill, *Death of Africa*, 101–2; Dawit, *Red Tears*, 351–55. Resettlement was already under way in late 1984. For informed accounts of resettlement, see Alula Pankhurst, *Resettlement and Famine in Ethiopia: The Villagers' Experience* (Manchester: Manchester University Press, 1992); and Alula Pankhurst and François Piguet, eds., *People, Space and the State: Migration, Resettlement and Displacement in Ethiopia* (Addis Ababa: Ethiopian Society of Sociologists, Social Workers and Anthropologists, 2004).

66. Pankhurst and Piguet, *People, Space and the State*, 10–11; Gill, *Year in the Death*, 142–52. It is frequently stated that ever since the expansion of the empire in the later nineteenth century, there had been a general southward drift of Ethiopian farmers, anticipating the formal programs of the 1970s and 1980s. The evidence for this is slight. The great majority of northern Ethiopians who settled in the south at the end

of the nineteenth century and on into the twentieth were soldiers who manned the garrisons, which maintained central government control over the areas. Again, in the years after 1945, the imperial government rewarded retired military and civilian officials with grants of land in the south. For this later program, see particularly John M. Cohen, "Ethiopia after Haile Selassie: The Government Land Factor," *African Affairs* 72, no. 289 (1973): 365–82; Charles W. McClellan, *State Transformation and National Integration: Gedeo and the Ethiopian Empire, 1895–1935* (East Lansing: African Studies Center, Michigan State University, 1988); and Sandra F. Joireman, *Property Rights and Political Development in Ethiopia and Eritrea* (Oxford: James Currey, 2000), chap. 5.

67. Jansson, Harris, and Penrose, *Ethiopian Famine*, 157. See also Dawit's account, *Red Tears*, chap. 7.

68. "Ethiopia Expects Crops to Be Best in Five Years," *Baltimore Sun*, November 17, 1986. The story was rerun by the *Christian Science Monitor* under the title "Ethiopia Aid Workers Shift Focus to Long Term," December 4, 1986.

69. The military story of the Darg's fall is ably told by Gebru, *The Ethiopian Revolution*. In 1987 the Darg held a constitutional referendum, which resulted in the creation of the Peoples' Democratic Republic of Ethiopia; Clapham, *Transformation and Continuity*, 92–100; Andargachew, *Ethiopian Revolution*, chap. 10, 265–98. The 1995 constitution reconstituted the country as a federation of ethnically determined units. For a sympathetic analysis, see Fasil Nahum, *Constitution for a Nation of Nations: The Ethiopian Prospect* (Lawrenceville, NJ: Red Sea Press, 1997). For discussions of Ethiopia and the Horn following the ethnic construction of the FDRE, see M. A. Mohamed Salih and John Markakis, *Ethnicity and the State in Eastern Africa* (Uppsala: Nordiska Afrikainstitutet, 1998); and Wendy James, Donald L. Donham, Eisei Kurimoto, and Alessandro Triulzi, *Remapping Ethiopia: Socialism and After* (Oxford: Ohio University Press, 2002).

70. For the early 1990s, see Webb and von Braun, *Famine and Food Security*, 21; for the more recent years, see Fassil, *Enough with Famines*, 28–30. For the revival of the strategy of resettlement, see Pankhurst and Piguet, *People, Space and the State.*

Chapter 3. Environment and Famine

1. Dessalegn Rahmato describes a similar experience: "When we first visited the province of Wollo in northeast Ethiopia in October 1986 to start the field-work for this study, what we saw was not exactly what we had been led to expect. The countryside appeared verdant and full of life, the harvest on many peasant plots was in good condition, and the people, young and old, bore no visible marks of having recently endured a traumatic experience." Dessalegn, *Famine and Survival Strategies*, 13.

2. See Allan Hoben, "Paradigms and Politics: The Cultural Construction of Environmental Policy in Ethiopia," *World Development* 23, no. 6 (1995): 1007–21.

3. For an informative view of the first five years of the SCRP, see "African Mountains and Highlands: Proceedings of an International Workshop organized by the Commission on Mountain Geoecology of the International Geographica Union (IGU) on behalf of the United Nations University, Addis Ababa, Ethiopia, 18–26 October 1986," a special issue of *Mountain Research and Development* 8, nos. 2/3 (May/August 1988). For the network of research stations, see M. Grunder, "Soil Conservation Research in

Ethiopia," *Mountain Research and Development* 8, nos. 2/3 (May/August 1988): 145–51; for major findings, see Hans Hurni, "Degradation and Conservation of the Resources in the Ethiopian Highlands," *Mountain Research and Development* 8, nos. 2/3 (May/August 1988): 123–30.

4. Hurni, "Degradation and Conservation."

5. Alemneh, *Environment, Famine and Politics*, 3. Alemneh cites M. Constable and members of the Ethiopian Highland Reclamation Study, "The Degradation of Resources and an Evaluation of Actions to Combat It," Working Paper no. 19, Ministry of Agriculture, Addis Ababa, December 1984, for the first statement of the 2010 prediction. See also Michael Constable and Deryke Belshaw, "The Ethiopian Highlands Reclamation Study: Major Findings and Recommendations," in *Towards a Food and Nutrition Strategy for Ethiopia: The Proceedings of the National Workshop on Food Strategies for Ethiopia, Held at Alemaya University of Agriculture, 8–12 December 1986* (Addis Ababa: People's Democratic Republic of Ethiopia, Office of the National Committee for Central Planning, 1989), 153.

6. I have been unable to consult the EHRS reports, which are scarce outside of Rome and Addis Ababa. In addition to the summary account by Constable and Belshaw, I have seen Michael Constable, "The Degradation of Resources and Evaluation of Actions to Combat It"; Getachew Yoseph and Mekonnen Tadesse, "Present and Projected Population"; and Yeraswork Admassie, Mulugetta Abebe, Markos Ezra, and J. Gay, "Report on the Sociological Survey and Sociological Considerations in Preparing a Development Strategy" (Institute of Development Research, Addis Ababa University, December 1983).

7. Constable and Belshaw, "Major Findings and Recommendations," 144–46.

8. Ibid., 147–48.

9. Ibid., 148–53. See chapters 4 and 5 for further evaluation of some of these claims.

10. Dessalegn Rahmato, "Littering the Landscape: Environmental Policy in Northeast Ethiopia," in *African Savannas: Global Narratives and Local Knowledge of Environmental Change*, ed. Thomas J. Bassett and Donald Crummey (Oxford: James Currey, 2003), 212.

11. For a trenchant critique, see J. P. Sutcliffe, "Soil Conservation and Land Tenure in Highland Ethiopia," *Ethiopian Journal of Development Research* 17, no. 1 (April 1995): 63–88. For a sympathetic evaluation of the soil management techniques of some Wallo farmers, see Belay Tegene, "Indigenous Soil Knowledge and Fertility Management Practices of the South Wallo Highlands," *Journal of Ethiopian Studies* 31, no. 1 (1998): 123–58; and Belay Tegene, "Combining Land Capability Evaluation, Geographic Information Systems, and Indigenous Technologies for Soil Conservation in Northern Ethiopia," *East African Social Science Research Review* 19, no. 2 (June 2003): 23–53.

12. For a fuller statement of this critique, see chapter 5.

13. To be sure, not all effective strategies were optimally deployed, and there were existing practices, like mono-cropping, multiple plowing of the land in preparation for planting smaller seeds like the highly valued *tef*, and cultivating steep slopes, which did contribute to erosion. Dessalegn, *Famine and Survival Strategies*, 52. Hoben argues that

both pre- and postrevolutionary land tenure arrangements in the plow-cultivated highlands discouraged long-term investments by farmers/users or landlords; Hoben, "Paradigms and Politics," 1009–10.

14. Quotation from Hurni, "Degradation and Conservation," 128.

15. The numbers are not official, to be sure, but consistently reported by different sources: Hurni, "Degradation and Conservation," 128. See also Yeraswork, *Twenty Years to Nowhere*, xxiii–xxiv; and Michael Stahl, "Environmental Degradation and Political Constraints in Ethiopia," *Disasters* 14, no. 2 (1990): 143–44. Stahl had already published a piece with the same title but somewhat different text: "Environmental Degradation and Political Constraints in Ethiopia," in *Ecology and Politics: Environmental Stress and Security*, ed. Hjort af Ornäs and M. A. Mohamed Salih (Uppsala: Scandinavian Institute of African Studies, 1989), 181–96.

16. Dessalegn, "Littering the Landscape," 216.

17. Ibid., 216–17.

18. Hansson, *The Ethiopian Economy*, 90–94.

19. Yeraswork, *Twenty Years to Nowhere*, 72–79. For the quotations, see 74 and 76. The Dessie archives document the destruction of controlled hillside plantations throughout Wallo. Bahru Zewde, "Forests and Forest Management in Wallo in Historical Perspective," *Journal of Ethiopian Studies* 31, no. 1 (1998): 106–7. Mesfin Wolde Mariam documented farmer frustration with the afforestation program in his research carried out 1986–88. Mesfin Wolde Mariam, *Suffering under God's Environment: A Vertical Study of the Predicament of Peasants in North-Central Ethiopia* (Berne: African Mountains Association and Geographica Bernensia, 1991), 44–45.

20. Hoben, "Paradigms and Politics," 1016, referring to the findings of Karl Herweg, a participant in the SCRP.

21. Jason W. Clay and Bonnie K. Holcomb, *Politics and the Ethiopian Famine 1984–1985* (Cambridge: Cultural Survival, 1986), 87; Pankhurst, *Resettlement and Famine in Ethiopia*, 66. See also Robert D. Kaplan, *Surrender or Starve: The Wars behind the Famine* (Boulder, CO: Westview Press, 1988), for an account similar to that of Clay and Holcomb. Some Wallo informants, interviewed in 1997, claimed there was no coercion; others said there had been. These interviews are the basis for chapter 5.

22. The only figures available are those of the RRC, which claimed a total of 593,000 people resettled, of whom 373,000 (61.0 percent of the total) came from Wallo, 107,000 (18.1 percent) came from Shawa, and 90,000 (15.2 percent) came from Tigray. The 6 percent unaccounted for here came from Kaffa, Gondar, and Gojjam. See Pankhurst, *Resettlement and Famine in Ethiopia*, 56, table 1. The 373,000 constituted 11 percent of Wallo's rural population at the time.

23. Constable and Belshaw, "The Ethiopian Highlands Reclamation Study," 169.

24. Alemneh, *Environment, Famine, and Politics*, 55–56 and 84–85; Yilma Seleshi and Ulrich Zanke, "Recent Changes in Rainfall and Rainy Days in Ethiopia," *International Journal of Climatology* 24 (2004): 973–83.

25. In addition to Yilma and Zanke, "Recent Changes," see Woldeamlak Bewket and Declan Conway, "A Note on the Temporal and Spatial Variability of Rainfall in the Drought-Prone Amhara Region of Ethiopia," *International Journal of Climatology* 27 (2007): 1467–77; Zewde Tessema Segele and Peter J. Lamb, "Interannual Variability

of Growing Season over Drought-Prone Areas of Ethiopia," *CLIVAR Exchanges* 8, nos. 2/3 (2003): 36–39; and Elisabeth Meze-Hausken, "Contrasting Climate Variability and Meteorological Drought with Perceived Drought and Climate Change in Northern Ethiopia," *Climate Research* 27 (2004): 19–31. The Ethiopian government started to collect national meteorological data in 1946, primarily for locust control and agriculture. The expansion of civil aviation in the early 1950s led to wider collection. The National Meteorological Services Agency was not established until December 1980. See http://www.ethiomet.gov.et/about/history (accessed May 11, 2012).

26. D. Conway and E. Lisa F. Schipper, "Adaptation to Climate Change in Africa: Challenges and Opportunities Identified from Ethiopia," *Global Environmental Change* 21 (2011): 227.

27. See, inter alia, Michael H. Glantz and Workineh Degefu, "Drought Issues for the 1990s," in *Climate Change: Science, Impacts and Policy: Proceedings of the Second World Climate Conference*, ed. J. Jäger and J. L. Ferguson (Cambridge: Cambridge University Press, 1991), 253–63.

28. Workineh, "Some Aspects of Meteorological Drought in Ethiopia," 25.

29. Woldeamlak and Conway, "Temporal and Spatial Variability," 1471.

30. D. Conway, "Some Aspects of Climate Variability in the North East Ethiopian Highlands—Wollo and Tigray," *SINET: An Ethiopian Journal of Science* 23, no. 2 (2000): 154; Eugene M. Rasmusson, "Global Climate Change and Variability: Effects on Drought and Desertification in Africa," in *Drought and Hunger in Africa: Denying Famine a Future*, ed. Michael H. Glantz (Cambridge: Cambridge University Press, 1987), 5.

31. For no statistically significant trends in rainfall variability, see Woldeamlak and Conway, "Spatial and Temporal Variability," 1468; and Yilma and Zanke, "Recent Changes," 973; for increasing rainfall at Dessie and Lalibäla, see Woldeamlak and Conway, "Spatial and Temporal Variability," 1472–73, 1476; and for variability in general, see Conway and Schipper, "Adaptation to Climate Change in Africa," 231, correcting a World Bank study that claimed variability ranging across ±60 percent of the mean.

32. Woldeamlak and Conway, "Temporal and Spatial Variability," 1467. For Alemneh's understanding, see *Environment, Famine, and Politics*, 35–37.

33. United Nations Environment Programme, *Report of the United Nations Conference on Desertification, 29 August–9 September 1977* (Nairobi: United Nations Environment Programme, 1977), 3. Lester Brown, director of the environmental organization Worldwatch Institute, was an articulate advocate of this position. See the discussion in chapter 2; Brown and Wolfe, *Reversing Africa's Decline*, 3. My discussion here draws heavily on Bassett and Crummey, "Contested Images, Contested Realities."

34. Richard H. Grove, *Green Imperialism: Colonial Expansion, Tropical Island Edens and the Origins of Environmentalism, 1600–1860* (Cambridge: Cambridge University Press, 1995).

35. For a brief account of the influence of French botanist Auguste Aubréville, see Bassett and Crummey, "Contest Images, Contested Realities," 10–12.

36. U. Helldén, "Desertification—Time for an Assessment?," *Ambio* 20, no. 8 (1991): 372–83; C. Tucket and H. D. Newcomb, "Expansion and Contraction of the Sahara Desert from 1980 to 1990," *Science* 253 (1991): 299–301. See also Michael Mortimer,

Adapting to Drought: Farmers, Famines and Desertification in West Africai (Cambridge: Cambridge University Press, 1987), 31–33.

37. Workineh, "Some Aspects of Meteorological Drought," 31–33. Workineh's account is generally paralleled by Kumar, who pays particular attention to FAO special reports on agricultural production. Kumar, *Ethiopian Famines*, 31ff.

38. Conway, "Some Aspects of Climate Variability," 158.

39. Alemneh, *Environment, Famine, and Politics*, 56, 57, and 84, tables 3.3, 3.4, and 4.2; Woldeamlak and Conway, "Temporal and Spatial Variability of Rainfall," 1470ff.

40. Woldeamlak and Conway, "Temporal and Spatial Variability of Rainfall," 1474.

41. Jansson, Harris, and Penrose, *The Ethiopian Famine*, 134–35. See also Kumar, *Ethiopian Famines*, 31ff.

42. De Waal, *Famine Crimes*, 115. His climate data are those reported by Alemneh, *Environment, Famine, and Politics*, 56, 57, and 84. He also claims that "Ethiopian statistics in the early 1980s cannot be taken at face value" (112), specifically how those pertaining to "rainfall, food production, prices and famine affected people" (114).

43. Zewde and Lamb, "Interannual Variability of Growing Season," 37, fig. 2, and 38.

44. Dessalegn, *Famine and Survival Strategies*, 105–6.

45. Woldeamlak and Conway, "Temporal and Spatial Variability," 1467, citing J. von Braun, *A Policy Agenda for Famine Prevention in Africa*, Food Policy Statement no. 13 (Washington, DC: International Food Policy Research Institute, 1991).

46. Glantz and Workineh, "Drought Issues for the 1990s." See also M. H. Glantz, ed., *Drought Follows the Plow* (Cambridge: Cambridge University Press, 1994). For a case from the United States, see Donald Worster, *Dust Bowl: The Southern Plains in the 1930s* (Oxford: Oxford University Press, 1979).

47. P. Stähli, "Changes in Settlement and Land-Use in Simien, Ethiopia, Especially from 1954 to 1976," in *Simien Mountains: Ethiopia*, vol. 1, *Cartography and Its Application for Geographical and Ecological Problems*, ed. B. Messerli and K. Aerni (Bern: Geographishches Institut der Universität Bern, 1978), 33–70.

48. James McCann, "Ethiopia," in *Drought Follows the Plow*, ed. M. H. Glantz (Cambridge: Cambridge University Press, 1994). See also James McCann, "The Social Impact of Drought in Ethiopia: Oxen, Households, and Some Implications for Rehabilitation," in *Drought and Hunger in Africa: Denying Famine a Future*, ed. Michael Glantz (Cambridge: Cambridge University Press, 1987), 245–67.

49. Meze-Hausken, "Contrasting Climate Variability," 20.

50. For an authoritative account of these wars based on Ethiopian archival materials, see Gebru, *The Ethiopian Revolution*. For briefer, also balanced accounts, see Bahru, *A History of Modern Ethiopia*, 257–62; Clapham, *Transformation and Continuity*, 201–14; and Keller, *Revolutionary Ethiopia*, 202–12.

51. See here Iyob, *The Eritrean Struggle for Independence*. From 1980 onward, Dawit Wolde Giorgis, a member of the central committee of the ruling party, was the party representative in Eritrea and hence in charge of the civilian side of the campaigns there. In 1983 he succeeded Shimellis Adugna as head of the RRC. Dawit, *Red Tears*, chap. 3. For more on Dawit, see chapter 2.

52. The leading study of the TPLF is Young, *Peasant Revolution in Ethiopia*.

53. See Andargachew, *The Ethiopian Revolution*, chaps. 9, 10; and Clapham, *Transformation and Continuity*, 77–100.

54. Gebru, *Ethiopian Revolution*, passim. For a lucid and succinct account, see Bahru, *Modern Ethiopia*, 264–68.

55. Keller, "Drought, War and the Politics of Famine," 615.

56. De Waal, *Famine Crimes*, 115–21. The book was copublished by African Rights and the International African Institute. For another exploration of the relations between the famine and war in Ethiopia, see Kaplan, *Surrender or Starve*. Kaplan is persuasive about the how the wars in Eritrea and Tigray exacerbated famine conditions and the Ethiopian government manipulated relief aid for political advantage. He never explicitly addresses the war as a causal factor of the famine. The earliest claim that military actions by the Ethiopian government were creating famine was made in a press release by the TPLF in Khartoum in 1980. *The Guardian*, October 22, 1980.

57. De Waal, *Famine Crimes*, 117–20. Keller writes of the systematic bombing of livestock, farms and food stores in Eritrea and Tigray: "Drought, War and the Politics of Famine," 620. De Waal is dismissive of the accounts by Jansson, Harris, and Penrose, *The Ethiopian Famine*; and by Dawit, *Red Tears*. As for the manipulation of relief programs, both the EPLF and the TPLF have been charged with attempting to do so for political ends. See the 1983 claim, in a story filed from Asmara in Eritrea, that "both the central authorities in Addis Ababa and the opposition seek political advance from international relief." *Christian Science Monitor*, October 25, 1983.

58. In April 1983 the Ethiopian RRC, then led by Shimellis Adugna, a man of noted integrity, claimed 2 million people at risk from drought in Gondar, Wallo, and Tigray. *Christian Science Monitor*, April 26, 1983. Already in 1983 several stories were filed from relief camps in Ebennat and Zwi Hamusit in eastern Gondar province. *Baltimore Sun*, May 2, 1983; *Washington Post*, June 27, 1983. Some of the people in these camps may have come from western Wallo, but the stories imply that most came from Gondar province and do not mention military operations in Tigray as a factor. Koram was the site where Kenyan journalist Mohammed Amin recorded the searing images, which, as discussed in chapter 2, were broadcast on British television at the end of October 1984 and propelled the famine to the forefront of international attention. Gill, *A Year in the Death*. Gill frames his book with an account of the Koram relief camp and sets out his own disturbing images in *Death of Africa*, especially chap. 8.

59. Gebru, *Ethiopian Revolution*, 137.

60. Gebru, *Ethiopian Revolution*, 96, 129, 131–33: 300,000–500,000 civilians, 400,000 Ethiopian combat personnel (as estimated by the Ministry of National Defense), 65,000–70,000 Eritrean combat personnel, and 50,000 Tigrayan combat personnel. The wars were also a great drain on the country's financial resources. They consumed some 50–60 percent of the government's operational budget and contributed over half of the country's foreign debt of US$ 9 million.

61. Lionel Cliffe, "The Impact of War and the Response to It in Different Agrarian Systems in Eritrea," *Development and Change* 20, no. 3 (1989), 373–400.

62. Clay and Holcomb, *Politics and the Ethiopian Famine*, 61–69.

63. Abenet Belete, John L. Dillon, and Frank M. Anderson, "Development of Agriculture in Ethiopia since the 1975 Land Reform," *Agricultural Economics* 6 (1991): 159.

64. Clapham, *Transformation and Continuity*, 166.

65. Ibid., 124, 161. See also Dessalegn, *Agrarian Reform in Ethiopia*, 67–68. Clapham and Dessalegn draw their information from national sources. For a congruent account resting on data from one peasant association in Wallo, a little to the north of Tahuladare, see Harald Aspen, *Competition and Co-operation: North Ethiopian Peasant Households and their Resource Base*, Working Paper on Ethiopian Development no. 7 (Trondheim: Trondheim Universitetet Centre for Environment and Development, 1993), 85–90.

66. The network of checkpoints, which de Waal describes for Tigray in the run-up to the famine, constrained grain movement within the province, applied over a much broader area. Quite possibly its application in Tigray was particularly strict, but insofar as I was able to observe myself in Gojjam, Gondar, and parts of northern Shawa, it reached every part of the country where the government had a presence. De Waal, *Famine Crimes*, 119–20.

67. Dessalegn notes that government officials throughout northeast Ethiopia wanted to limit farmer access to markets on the grounds that they were wasting time there. Dessalegn, *Famine and Survival*, 207.

68. Dessalegn, *Famine and Survival*, 222. See also Mesfin, *Suffering under God's Environment*, 148, for an increase in cash payments by area farmers in 1983–84, and 185 for the growth in the scale of rural taxation over 1958–86.

69. Population figure from http://www.unicef.org/infobycountry/ethiopia_statistics (accessed April 13, 2012). This annual rate of growth is for the period 1970–90. Mesfin cites the CSO for an earlier growth rate of 2.9 percent each year. Mesfin, *Suffering under God's Environment*, 59.

70. Clay and Holcomb, *Politics and the Ethiopian Famine*, 194. For the quotation from the 1975 proclamation, see Clapham, *Transformation and Continuity*, 171.

71. Keller, *Revolutionary Ethiopia*, 260, table 9.1.

72. Clay and Holcomb, *Politics and the Ethiopian Famine*, 192: "The intensification of resettlement which required indigenous people to supply land, houses, food, equipment and services to arriving settlers, introduced famine where it had not been known recently." The best studies of resettlement are Pankhurst, *Resettlement and Famine in Ethiopia*; and Pankhurst and Piguet, *People, Space and the State*.

73. Clapham, *Transformation and Continuity*, 172. See ibid., 174–79, for Clapham's discussion of cooperatives. See also Keller, *Revolutionary Ethiopia*, 227–30; and Paul Henze, *The Primacy of Economics for the Future of the Horn of Africa* (Santa Monica, CA: Rand, 1992), 4–5.

74. Clapham, *Transformation and Continuity*, 161–68; for the AMC, 168–71. See also Keller, *Revolutionary Ethiopia*, 257–62.

75. Yeraswork, *Twenty Years to Nowhere*; Dessalegn, "Littering the Landscape." See also Hoben, "Paradigms and Politics." For a fuller account of the critiques, see chapter 4.

76. J. Nyssen, A. Frankl, R. N. Munro, P. Billi, and M. Haile, "Digital Photographic Archives for Environmental and Historical Studies: An Example from Ethiopia," *Scottish Geographical Journal* 126, no. 3 (2010): 185–207; J. Nyssen, M. Haile, J. Naudts, R. N. Munro, J. Poesen, J. Moeyersons, A. Frankl, J. Deckers, and R. Pankhurst,

"Desertification? Northern Ethiopia Re-photographed after 140 Years," *Science of the Total Environment* 407 (2009): 2749–55.

Chapter 4. Farmers Cope

1. I have not had access to all the reports produced by the EHRS. My account relies primarily on the summary by Constable and Belshaw, "The Ethiopian Highlands Reclamation Study," 142–79. For a particularly eloquent call to incorporate local perceptions into the analyses of landscape stress and local participation in the planning of interventions, see Yeraswork et al., "Report on the Sociological Survey."

2. Two early collections laid out the parameters of the existing state of knowledge: M. H. Glantz, ed., *The Politics of Natural Disaster: The Case of Sahel Drought* (New York: Praeger 1976), which, as the title makes clear, was primarily directed to the West African Sahel; and Abdul Mejid, ed., *Drought and Famine in Ethiopia* (London: International African Institute, 1976). Journalists made some contributions of lasting value, notably Shepherd, *The Politics of Starvation.*

3. Rivers et al., "Lessons for Epidemiology from the Ethiopian Famines"; and Julius Holt and John Seaman, "Markets and Famines in the Third World," *Disasters* 4, no. 3 (1980): 283–97. See also Holt and Seaman, "The Scope of the Drought"; J. B. Mason, R. W. Hay, J. Holt, J. Seaman, and M. R. Bowden, "Nutritional Lessons from the Ethiopian Drought," *Nature* 248 (April 19, 1974): 646–50; and Julius Holt, John Seaman, and John Rivers, "Famine Revisited," *Nature* 255 (May 15, 1975): 180–81. The work by Seaman and Holt on the 1974–75 famine in Ethiopia was foundational for careers dedicated to international development and disaster response and continuing involvement with Ethiopia. Seaman was cofounder and first editor of the journal *Disasters.* Holt is attached to the Food Economy Group, which helped produce the authoritative and informational *Atlas of Ethiopian Livelihoods* (2011), which describes itself as "a joint Government of Ethiopia and USAID Ethiopia document." There were still other researchers who produced work of lasting value before 1980. See, for example, Mehari Gebre-Medhin and Bo Vahlquist, "Famine in Ethiopia—The Period 1973–75," *Nutrition Reviews* 35, no. 8 (August 1977): 194–202; and Shewandagn Belete, Mehari Gebre-Mehin, Bantirou Hailemariam, Mario Maffi, Bo Vahlquist, and Zewdie Wolde-Gebriel, "Study of Shelter Population in the Wollo Region," *Journal of Tropical Pediatrics and Environmental Child Health* 23, no. 1 (February 1977), Monograph 48, "Famine in Ethiopia," ed. Bo Vahlquist, 15–22.

4. Stephen Devereux, "Entitlements, Availability and Famine: A Revisionist View of Wollo, 1972–74," *Food Policy* (August 1988): 270–82.

5. Sen, *Poverty and Famines*, 93–96. For the movement of food out of Wallo, he cites Holt and Seaman, "The Scope of the Drought."

6. Sen, *Poverty and Famines*, 98–99.

7. Seaman and Holt, "Markets and Famines," 286.

8. Ibid., 293.

9. For an important exception, see Shiferaw Bekele, *An Economic History of Modern Ethiopia*, vol. 1, *The Imperial Era, 1941–74* (Dakar: CODESRIA, 1995). Only one volume appeared.

10. Seaman and Holt, "Markets and Famines," 286.

11. Rivers et al., "Lessons for Epidemiology," 347. The 1974 reference is to an unpublished report Seaman and Holt submitted to the Save the Children Fund.

12. Seaman and Holt, "Market and Famines," 292–93.

13. Devereux, "Entitlements, Availability and Famine," 276.

14. Cutler, "Famine Forecasting." While recognizing the integrity of the early data it generated, Alex de Waal has attacked all the RRC data surrounding famine in the 1980s on food production, food shortages, numbers of people affected, and so on, as exaggerated and fundamentally distorted by overriding political imperatives. De Waal, *Famine Crimes*, 111–13.

15. Cutler, "Famine Forecasting," 52–53. For an argument, pertaining to a later date, 1990, that there was no evidence of a drought-induced rise in the price of cereals, see Endale, *Rural Markets*, 19–21.

16. Devereux, "Entitlements, Availability and Famine." See also Kumar, *Ethiopian Famines 1973–1985*.

17. McCann, *From Poverty to Famine*, 33–34. I discuss the photographic evidence from the Simien Mountains, generated by B. Messerli and R. Aerli, in chapter 4 and in Donald Crummey, "Exploring Landscape Change in Ethiopia: Evidence from Imaging and its Interpretation," in *Research in Ethiopian Studies: Selected Papers of the 16th International Conference of Ethiopian Studies, Trondheim July 2007*, ed. Harald Aspen, Birhanu Teferra, Shiferaw Bekele, and Svein Ege (Wiesbaden: Harrassowitz, 2010), 93–105.

18. McCann, *From Poverty to Famine*, 58–60. This is the position advanced by James Scott, *Seeing Like a State: How Certain Schemes to Improve the Human Condition Have Failed* (New Haven, CT: Yale University Press, 1998). My informants in Dambeya in 1988 expressed a decided preference for fixed payments, because they found these were almost always lower than proportional ones would have been.

19. McCann, *From Poverty to Famine*, 140; see also 72 and 133. For the general discussion on increased pressures on farmers in the 1920s and 1930s, see *From Poverty to Famine*, chap. 6 "The State and Rural Society in Northern Wallo: Households and the Imperial Order," 127–45.

20. McCann, *From Poverty to Famine*, 30–33.

21. Ibid., 34, 174, and 187. For his revisionist account, see James McCann, *People of the Plow: An Agricultural History of Ethiopia, 1800–1990* (Madison: University of Wisconsin Press, 1995), chap. 4, especially 119–24, and chap. 5. He also explored this material explored in James McCann, "The Plow and the Forest: Narratives of Deforestation in Ethiopia, 1840–1992," *Environmental History* 2, no. 2 (1997): 138–59.

22. McCann, *From Poverty to Famine*, 206. McCann has particularly emphasized oxen—as opposed to land—as the critical factor in highland Ethiopian agriculture. McCann, "The Social Impact of Famine in Ethiopia."

23. Mesfin, *Rural Vulnerability to Famine in Ethiopia*.

24. Ibid., 165–74. The very high incidence of famine as reported by Mesfin has not received the attention or scrutiny that it deserves. Few authorities have followed him here.

25. Ibid., 178.

26. Mesfin, *Suffering under God's Environment*.

27. The social, economic, and geographical information tabulated in Mesfin's chapters 2–8 is presented in 200-meter intervals; the larger agroecological categories are discussed in chapter 9 generally in 400-meter intervals.

28. Mesfin, *Suffering under God's Environment*, 42ff.

29. Ibid., 187. The general discussion is on 177–87.

30. For Mesfin's discussion of markets and his argument that they exploit farmer vulnerability, see *Suffering under God's Environment*, chap. 6, "Peasant Trade," 123–37.

31. For an account that emphasizes farmer capacity for change and innovation, see Donald Crummey and Alex Winter-Nelson, "Farmer Tree-Planting in Wallo, Ethiopia, 1937–1997," in *African Savannas: Global Narratives and Local Knowledge of Environmental Change*, ed. Thomas J. Bassett and Donald Crummey (Oxford: James Currey, 2003), 91–120. See also chapter 6 in this volume. For an account that, like Mesfin, finds the rigidity and incapacity to develop one of the principal factors making peasant society vulnerable to famine, see Adhana, "History of Selected Famines," 606.

32. See the section "Markets Revisited" below.

33. Alemneh, *Environment, Famine, and Politics in Ethiopia.*

34. Ibid., 3, citing M. Constable and Members of the Ethiopian Highland Reclamation Study, *The Degradation of Resources and an Evaluation of Actions to Combat it* (Addis Ababa: Ministry of Agriculture, December 1984), Working Paper no. 19.

35. Ibid., 19 and 35ff.

36. Ibid., 44. For Alemneh's general discussion of deforestation, see 35–46. Alemneh's findings on farmer tree planting are discussed in chapter 7.

37. Ibid., 54. For the general discussion of soil erosion, see 50–55.

38. The book appeared earlier as a report.

39. Tesfaye Wolde Medhin, "Highland Farmers and the "Modernizing" State in Ethiopia: Conjunctures and disjunctures" (PhD diss., University of Illinois at Urbana-Champaign, 2004), chap. 4, 116ff.

40. Dessalegn, *Famine and Survival Strategies*, 21.

41. Dessalegn (*Famine and Survival Strategies*, 25) denies the analytical utility of the term "rural proletariat," although he does implicitly allow for its existence. My own research in Wallo and in Dambeya, carried out over twenty years after the great land reform, found vestiges of a rural proletariat but accords with Dessalegn's judgment that it was a "puny force." Of 120 informants, perhaps 2 might be said to fit the category.

42. Dessalegn, *Famine and Survival Strategies*, 69–70.

43. Alex Winter-Nelson, "Rural Taxation in Ethiopia, 1981–1989: A Policy Analysis Matrix Assessment for Net Producers and Net Consumers," *Food Policy* 22 (1997): 419–32. The years 1983–84, of course, were by no means typical but this was when the famine hit Wallo very hard.

44. Julius Holt and Mark Lawrence, *Making Ends Meet: A Survey of the Food Economy of the Ethiopian North-East Highlands* (London: Save the Children UK, 1993), 144.

45. For earlier remarks on the Agricultural Marketing Corporation, see chapter 3.

46. Dessalegn, *Famine and Survival*, 94–95.

47. Ibid., 143ff.

48. Dessalegn, *Famine and Survival*, 149–50.

49. Adhana, "History of Selected Famines."

50. Adhana, "History of Selected Famines," 67–68: 30,000 tons = 15,000 (sorghum) + 10,000 (wheat) + 5,000 (maize). Wallo's position in the national grain market suffered heavily, apparently because of the famines. In the early 1990s a study based on interviews with grain traders and brokers in Addis Ababa and Dessie claimed that the volume of trade was negligible and seriously in deficit. In the first four months of 1994, Addis Ababa imported 176,000 metric tons of grain, of which 1,000 (0.6 percent) came from Dessie. Of the 92,000 metric tons of grain Addis Ababa exported to the provinces, 1,000 (1.1 percent) went to Dessie. Taking into account other transfers, Dessie's grain trade was 8,000 tons in deficit.

51. Adhana, "History of Selected Famines," 78–118.

52. Ibid., 119–62.

53. For a definitive account of the rebellion, see Gebru Tareke, *Power and Protest: Peasant Revolts in the Twentieth Century* (Cambridge: Cambridge University Press, 1991).

54. Ibid., 163–310.

55. Ibid., 324.

56. Ibid., 438.

57. Ibid., 376–90.

58. These taxes and the broader administrative organization of which they were a part have been discussed by a number of authorities. Adhana, "History of Selected Famines," 553–58: M. Perham, *The Government of Ethiopia* (London: Faber and Faber, 1948); Bahru, *A History of Modern Ethiopia*; Markakis, *Ethiopia*; Gilkes, *The Dying Lion*.

59. Winter-Nelson, "Rural Taxation in Ethiopia."

60. Holt and Lawrence, *Making Ends Meet*, 34.

61. Berhanu Abegaz, "Persistent Stasis in a Tributary Mode of Production: The Peasant Economy of Ethiopia," *Journal of Agrarian Change* 5, no. 3 (July 2005): 299–333, quoted on 299.

62. See, for example, Crummey and Winter-Nelson, "Farmer Tree-Planting."

63. Markos Ezra, "Demographic Responses to Ecological Degradation and Food Insecurity. Drought Prone Areas in Northern Ethiopia" (PhD diss., Royal University of Groningen, 1997). Markos had already contributed to the EHRS. Yeraswork et al., *Report on the Sociological Survey and Sociological Considerations*.

64. Markos, "Demographic Responses," 40.

65. One solution, practiced in north Shawa, was sending their cattle to neighboring Oromo peasant associations. However, the fees charged by the Oromo were felt to have been exorbitant.

66. Markos, "Demographic Responses," 107–19.

67. Ibid., 133, 135–37.

68. Something appears slightly amiss here. On p. 245 of Markos, "Demographic Responses," these figures are given as 2.2 percent for rural and 4.1 percent for urban, but on p. 247 we are simply told that the overall national growth rate was 2.2 percent. Admittedly the urban population was much smaller than the rural population, but not such as to have had no effect on the overall figure.

69. Markos, "Demographic Responses," 342–45. Markos, agreeing with other scholars that tenure insecurity is a real problem, cites the mismanagement of community-owned resources as an example. These seem to me to be two separate problems; individual tenure security may indeed discourage investment, but the rules and conventions governing the management of common lands are distinct.

70. See the discussion on marriage and divorce in chapter 6.

71. Yeraswork Admassie, Mulugeta Abebe, Markos Ezra, and J. Gay, with the assistance of Simon Adebo, *Ethiopian Highlands Reclamation Study: Report on the Sociological Survey and Sociological Considerations in Preparing a Development Strategy* (Addis Ababa, December 1983), Ethiopian Highlands Reclamation Study, Working Paper no. 4.

72. Yeraswork, *Twenty Years*, 290ff.

73. Ibid., xxii–xxiii. By 1990 the program could claim 1 million km of soil and stone bunds, more than 470,000 hillside terraces, and 300,000 hectares reforested, much of it through hillside enclosures. See also the discussion in chapter 3, in the section "Farming Practice."

74. Yeraswork, *Twenty Years*, 166–67.

75. Ibid., xxiv–xxix.

76. Ibid., chap. 3, "Outcome of the Attempt at Inducing Conservation and Afforestation Innovations: As if Written on Water."

77. Ibid., 76. "Unrestrained," where Yeraswork gives "unguarded" for the Amharic *takalakay yalelaw*.

78. Ibid., 93.

79. Ibid., 167. Where farmers plant trees is an issue here. Most often they planted them around their households and gardens. Less commonly, they planted them at various places across their holdings. Finally and most ambitiously, some farmers converted grain fields to tree lots. Yeraswork cites a study by Markos Ezra and Kassahun Berhanu that reports on tree planting in the Wallo *awraja*s of Ambassel and Wara-Ilu—71.7 percent of respondents planted trees around their houses, 11.7 percent planted around their farm plots, and 8.3 percent planted trees interspersed with crops. Markos Ezra and Kassahun Berhanu, *A Review of the Community Forestry Programme and an Evaluation of Its Achievements: A Socio-Economic Survey* (Addis Ababa: Swedish International Development Agency, 1988).

80. The foregoing is based on Yeraswork, *Twenty Years*, 92–98. I take up farmer tree planting again in chapter 6.

81. Yeraswork, *Twenty Years*, chap. 8.

82. Dessalegn, "Littering the Landscape."

Chapter 5. Parameters of Landscape Change in Wallo

1. The following account draws on Crummey, "Exploring Landscape Change in Ethiopia," and on Donald Crummey, "Deforestation in Wallo: Process or Illusion?," *Journal of Ethiopian Studies* 36, no.1 (1998): 1–41.

2. Stähli, "Changes in Settlement"; P. Stähli and M. Zurbuchen, "Two Topographic Maps 1:25,000 of Simien, Ethiopia," in *Simien Mountains: Ethiopia*, vol. 1, *Cartography and Its Application for Geographical and Ecological Problems*, ed. B. Messerli

and K. Aerni (Bern: Geographishches Institut der Universität Bern, 1978), 11–31. I have not had access to the 1:25,000 topographic maps or a color-coded map, perhaps the same one Stähli refers to in his article. I have had access to related maps, two of them at 1:100,000 by Hans Hurni, *Climate and Dynamics of Altitudinal Belts from the Last Cold Period to the Present Day*, vol. 2 of *Simien Mountains: Ethiopia* (Bern: Geographishches Institut der Universität Bern, 1982), Geographica Bernensia G 13; along with another map, 1:50,000, also by Hans Hurni, "Harvesting Areas and Cultivation Belts in and around the Simien Mountains National Park—Ethiopia," n.d. The "Harvesting" map is based, topographically, on the Stähli and Zurbuchen 1:25,000 map and reports data collected in fieldwork work from 1975 to 1977.

3. Stähli, "Changes in Settlement," 51. Stähli cites Mesfin Wolde Mariam, *An Introductory Geography of Ethiopia* (Addis Ababa, 1972), as his source. A more common statement about deforestation is the claim that Ethiopia's forest cover has diminished from 40 percent to 4 percent within the twentieth century, which James McCann traces to Huffnagel, *Agriculture in Ethiopia*; see McCann "The Plow and the Forest." See a similar account by Alain Gascon, "La forêt perdue d'Éthiopie, un mythe contemporain," in Monique Chastanet, *Plantes et paysages d'Afrique: Une histoire à explorer* (Paris: Editions Karthala, Centre de recherches africaines, 1998), 390–98.

4. Stähli, "Changes in Settlement," 36. The accusation that local farmers lack understanding of ecological problems belittled generations of practical knowledge. Bassett and Crummey, "Contested Images, Contested Realities." For the distinction between local, practical knowledge and "scientific" knowledge, see Scott, *Seeing Like a State*.

5. Stähli, "Changes in Settlement," 69. While the subsequent occurrence of famine in 1984 was indeed catastrophic, it is unclear whether it was caused by indigenous agricultural practice.

6. Mary Tiffen, Michael Mortimore, and Francis Gichuki, *More People, Less Erosion: Environmental Recovery in Kenya* (London: John Wiley, 1994). Also using matched photographs to contribute to a revised view of African landscape change was Richard Rohde, "Nature, Cattle Thieves and Various Other Midnight Robbers: Images of People, Place and Landscape in Damaraland, Namibia" (PhD diss., Edinburgh University, 1997).

7. James Fairhead and Melissa Leach, *Misreading the African Landscape: Society and Ecology in a Forest-Savanna Mosaic* (Cambridge: Cambridge University Press, 1996).

8. McCann, *People of the Plow*, chap. 4, especially 119–24, and chap. 5. This material he also explored in "The Plow and the Forest."

9. Halvor Wøien, *Woody Plant Cover and Farming Compound Distribution on the Mafud Escarpment, Ethiopia: An Aerial Photo Interpretation of Changes 1957–1986*, Working Paper on Ethiopian Development no. 9 (Trondheim: University of Trondheim, Centre for Environment and Development, 1995).

10. This was Ritler's master's thesis, which ostensibly looked at records beginning in 1699 but was focused on the period from 1830 to 1865. For an abbreviated version, see Alfons Ritler, *Land Use, Forests and the Landscape of Ethiopia, 1699–1865: An Enquiry into the Historical Geography of Central-Northern Ethiopia*, Soil Conservation Research Programme Ethiopia, Research Report no. 38 (Bern: University of Bern, 1997). This he

followed with a PhD dissertation, "Wald, Landnutzung und Landschaft im Zentralen und Nördlichen Äthiopien 1865–1930" (Centre for Development and Environment, Bern, 2001), in two volumes, the second of which has a comprehensive presentation of his source material organized on a regional basis, which was published in an English translation, without the second volume, in 2003 as Alfons Ritler, *Forests, Land Use and Landscape in the Central and Northern Ethiopian Highlands, 1865 to 1930* (Bern: University of Berne, Institute of Geography, 2003), Publication A 19 in the African Studies Series of Geographica Bernensia.

11. So far, sources in Ethiopian languages have yielded little information about historic landscapes.

12. Ritler, *Forests, Land Use and Landscape*, xii.

13. Gascon, "La forêt perdue d'Éthiopie"; more recently, Alain Gascon, *Sure les hautes terres comme au ciel: Identités et territoires en Ethiopie* (Paris: Publications de la Sorbonne, 2006).

14. A. Gascon, "La forêt perdu," 387.

15. Boerma, "Seeing the Wood for the Trees," which anticipated Pauline A. Boerma, "Assessing Forest Cover Change in Eritrea—A Historical Perspective," *Mountain Research and Development* 25, no. 1 (2006): 41–47.

16. Dino Scheidegger, *Äthiopien rekonstruiert: Landschafts-, Landnutzungs- und Walddynamik im Hochland von Äthiopien zwischen 1930 und 1998 anhand von Photovergleichen* (Bern: Centre for Development and Environment, Geographisches Institut, 1999).

17. Robin S. Reid, Russell L. Kruska, Nyawira Muthui, Andualem Taye, Sara Wotton, Cathleen J. Wilson, and Wondyalew Mulatu, "Land-Use and Land-Cover Dynamics in Response to Changes in Climatic, Biological and Socio-Political Forces: The Case of Southwestern Ethiopia," *Landscape Ecology* 15 (2000): 339–55.

18. The authors discuss these factors with reference to a broad comparative literature. Reid et al., "Land-Use and Land-Cover Dynamics," 349–51.

19. For Wallo historical traditions, see D. Brielli, "Ricordi storici dei Uollo (con note di C. Conti Rossini)," in *Studi Etiopici*, ed. C. Conti Rossini (Rome: Istituto per l'Oriente, 1945), 78–110. A foundational work of later Wallo historiography was Zergaw Asfera, "Some Aspects of Historical Development in 'Amhara, Wallo,' *ca.* 1700–1815" (bachelor's thesis, Addis Ababa University, 1973). See also Hussein Ahmed, *Islam in Nineteenth-Century Wallo, Ethiopia: Revival and Reaction* (Leiden: Brill, 2001).

20. As always, Bahru Zewde is a good guide to these developments. Bahru, *History of Modern Ethiopia*, chaps. 2 and 3.

21. Marcus, *Haile Sellassie I*, 30.

22. Ritler, *Land Use, Forests and the Landscape*. Ritler's summary is on 38 and 39. Appendix II.5 (185–96) consists of verbatim reproductions of the landscape observations of these travelers.

23. Ritler, *Forests, Land Use and Landscape*, 90.

24. Richard Pankhurst, "The Genesis of Photography in Ethiopia and the Horn of Africa," *British Journal of Photography* (October 8, 15, 22, and 29, 1976): 878–82, 910–13, 933–35, and 952–53. The richest collections are in the British Army Museum's Ogilby Trust in London, the National Army Museum in London, and the library of

the Institution of the Royal Engineers, Kent. The Institute of Ethiopian Studies at Addis Ababa University also has several albums.

25. Ritler, *Forests, Land Use and Landscapes*, 94; Scheidigger, *Äthiopien rekonstruiert*, 45. Scheidigger reproduces later photos of the Ashangi area. For all of these, he provides matched photos taken in 1998. See also Richard Pankhurst and Denis Gerard, *Ethiopia Photographed: Historic Photographs of the Country and Its People Taken between 1867 and 1935* (London: Kegan Paul, 1996), 38–41, for additional photos from the Maqdala expedition, without subsequent rephotographs.

26. Nyssen et al., "Desertification?"

27. Alberto Sbacchi, *Ethiopia under Mussolini: Fascism and the Colonial Experience* (London: Zed Books, 1985); Alberto Sbacchi, *Il colonialism italiano in Etiopia 1936–1940* (Milan: U. Mursia, 1980); Haile Mariam Larebo, *The Building of an Empire: Italian Land Policy and Practice in Ethiopia 1935–1941* (Oxford: Clarendon Press, 1994). There is a run of consular reports from March 1933 to April 1934, which are primarily of interest for their insights into the administration of Wallo by Crown Prince Asfaw Wossen: Rome, Archivo Storico della Ministero Africano-Italiano, 54/6. I owe my access to these materials to the kindness of James McCann.

28. Bahru, "Forests and Forest Management." This issue of the journal was dedicated to papers produced by members of the research team funded by a grant from the John D. and Catherine T. MacArthur Foundation through the Center for African Studies of the University of Illinois.

29. Istituto Agronomico d'Oltremare, Florence: Milizia Nazionale Forestale, Comando XI Legione Milizia Forestale, *Relazione sulla Attività svolta dalla Costituzione (1° Gennaio 1937/XV) a tutto l'anno XVIo*; Governo Generale dell'A.O.I., Comando XI Legione Milizia Forestale, *Attività Forestali Negli Anni XVo e XVIo: Parte II: Allegati: Regolamentazione e studi di comprensori boschivi*. See the summary for Wallo, 249–51. For access to this and other material cited at the Istituto Agronomico d'Oltremare, I am indebted to Irma Taddia, who repeatedly helped with arrangements, and Marina Puccioni, of the institute's Documentation Center.

30. "Il nomadismo dell'agricoltura." Irrationality was one of the characteristics Italians commonly attributed to Ethiopian farmers. "Shifting agriculture" characterizes farming practices in southern and southwestern Ethiopia, but the ox-plow farmers of Wallo have had fixed fields for a long time. Italian sources appear to have had neither a firm nor a favorable grasp on Ethiopian farming practices.

31. Brielli, "Ricordi storici dei Uollo." For subsequent use, see a succession of bachelor's and master's theses submitted to the History Department, Addis Ababa University, among the earliest of which were Fekadu Begna, "A Tentative History of Wallo, 1855–1908" (bachlor's thesis, 1972); and Zergaw, "Some Aspects of Historical Development."

32. Alessandro Lusana, "L'Uollo Orientale," *Gli Annali dell'Africa Italiana* 2, no. 2 (1939): 477–515. Lusana also published an article for a more general audience: "L'Itinerari dell'A. O. I. La regione di Dessie," *L'Azione Coloniale* 10, no. 17 (April 25, 1940): 4—as did M. Corigliano, "Regioni Poco note dell'Africa O. I. La residenza del lago Haik," *Illustrazione Coloniale* 19, no. 3 (March 1941): 33–37.

33. Lusana, "L'Uollo Orientale," 487.

34. Ibid., 507.
35. Ibid., 508.
36. I am grateful to James McCann for drawing my attention to these photographs, to Irma Taddia for facilitating my access to them, and to Marina Puccioni and the staff of the Documentation Center of the Istituto Agronomico d'Oltremare for permission to use the collection.
37. Haile Mariam Larebo's account of Italian land policy in Ethiopia refers frequently to Maugini. Haile, *Building of an Empire*, 64 (n40), 69–71, 74, 76, 81, 153, 167, 194, 233, 239 (as indexed). See also Istituto Agronomico per l'Oltremare, *Armando Maugini nel centenario della nascita* (Florence, 1989); and Elena Maugini and Luca Fabbri, "Storia di una vita: Note biografiche di e su Armando Maugini," mimeograph in the Biblioteca della Facoltà di Agraria, Firenze, Fondo Maugini, 1948. For a copy of this last item, I am once again indebted to Irma Taddia.
38. Istituto Agronomico, *Armando Maugini*, 43. The camera he used in 1937 had a focal length of about 40–44 mm, and the 1940 camera had a focal length of 50 mm.
39. For some years in the late 1940s and through the 1960s, this road was one of the country's lifelines, connecting Addis Ababa to the Red Sea at the port of Assab, now in Eritrea, second in importance only to the railroad, which connects the capital to the port of Djibouti. The stretch from Kombolcha to Awash receded in importance following the completion of a shorter route along the Awash River that approached Addis Ababa from the southeast. Following Eritrean independence in 1993, and the wars with Ethiopia that followed, Assab became inaccessible to Ethiopia, and traffic on the road east from Kombolcha declined even further.
40. The 1940 photographs, some of which I have replicated, are valuable, as is the rest of this remarkable collection.
41. Archival references AOI, 2188, 1937, vol. 5: enlarged 7159/180 to 7183/204 and contact 50/8 to 52/3.
42. For the identification of Ethiopian vernacular plant names, I have relied on Wolde Michael Kelecha, *A Glossary of Ethiopian Plant Names*, 4th ed. (Addis Ababa: Addis Ababa University Press, 1987). I owe my copy to the kindness of the late Merid Wolde Aregay of the History Department, Addis Ababa University. My fieldwork was ably assisted by Merere Ejjeta.
43. Chuck Schaefer of Valparaiso University described the plain, which he knew as a youth, to have been heavily forested.
44. McCann, *People of the Plow*, 137–43. It is unclear what impact changing land tenure had on this process.
45. The figure of 40 percent forest cover to 4 percent probably preceded Huffnagel's book in the lectures of forestry professor F. von Breitenbach, who taught forestry to the earliest generation of Haile Selassie I University students. See F. von Breitenbach, "National Forestry Development Planning: A Feasibility and Priority Study on the Example of Ethiopia," *Ethiopian Forestry Review* 3 (1962): 43. While his statement was published later than Huffnagel's book, von Breitenbach's lectures on Ethiopia's forest cover were hypothetical, based purely on rainfall distribution, not surveys. See James C. McCann, *Green Land, Brown Land, Black Land: An Environmental History of Africa, 1800–1990* (Portsmouth, NH: Heinemann, 1999), 80–81.

46. Belay Tegene, "Land Cover/Land-Use Changes in the Derekolli Catchment of the South Welo Zone of Amhara Region, Ethiopia," *Eastern Africa Social Science Research Review* 18, no. 1 (January 2002): 1–20. See also Belay Tegene, "Potentials and Limitations of an Indigenous Structural Soil Conservation Technology of Welo, Ethiopia," *Eastern Africa Social Science Research Review* 14, no. 1 (January 1998): 1–18; and Belay, "Combining Land Capability Evaluation."

47. Girmay Kassa Welela, "GIS Based Analysis of Land Use/Land Cover, Land Degradation and Population Changes (A Study of Boru-Metero Area of South Wello, Amhara Region)" (master's thesis, Addis Ababa University, 2003). Girmay's thesis also included more closely focused studies within Boru Metero peasant association of intensely settled Boru Sellasé and wooded Gora. The larger context is of primary interest here.

48. Kebrom Tekle and Lars Hedlund, "Land Cover Changes Between 1958 and 1986 in Kalu District, Southern Wello, Ethiopia," *Mountain Research and Development* 20, no. 1 (February 2000): 42–51.

49. In a study of the Gido catchment area, which includes much if not all of the Gwobeya peasant association, Belay's analysis indicated a high area under cultivation of 73.7 percent of the total land surface, with 9.9 percent in shrub and planted forest, 5.0 percent in shrub grassland, 3.81 percent grassland, 2.4 percent in valley-rim vegtation, and 4.9 percent in hedges and private woodlots. With insights gained from fieldwork in 2000, he extracted the information from the 1986 aerial photograph. Belay "Combining Land Capability Evaluation." The article is primarily directed to soil erosion/conservation issues.

50. Gete Zeleke and Hans Hurni, "Implications of Land Use and Land Cover Dynamics for Mountain Resource Degradation in the Northwestern Ethiopian Highlands," *Mountain Research and Development* 22, no. 2 (May 2001): 184–91.

51. Belay, "Land-Cover/Land-Use Changes," 18.

52. Kebrom and Hedlund, "Land Cover Changes," 48.

53. Girmay defines his categories in table 1, 15, of "GIS Based Analysis of Land Use/Land Cover."

54. The importance of tree planting for land cover is explored by Woldeamlak Bewket with reference to a part of Gojjam adjacent to Dambacha Warada, the site of the Gete and Hurni study. Woldeamlak Bewket, "Household Level Tree Planting and Its Implications for Environmental Management in the Northwestern Highlands of Ethiopia: A Case Study in the Chemoga Watershed, Blue Nile Basin," *Land Degradation and Development* 14 (2003): 377–88.

55. Belay, "Land-Cover/Land-Use Changes," 14.

56. See the section "Farmer Tree Planting" in chapter 6.

Chapter 6. Farmers and Landscape Change in Wallo

1. Not long after we chose Gerba, the *warada* boundaries were changed and Gerba was no longer in Tahuladare. I have been able to confirm the change from Svein Ege, *South Wälo 1:100,000: Topographic and Administrative Map of South Wälo Zone, Amhara, Ethiopia*. This invaluable map is available online at http://www.svt.ntnu.no/ethiopia/c16xSouthW%EloMapFull.pdf.

2. Thomas J. Bassett and Donald Crummey, eds., *African Savannas: Global Narratives and Local Knowledge of Environmental Change* (Oxford: James Currey, 2003). Preliminary findings of the Ethiopian team were published in a special issue of the *Journal of Ethiopian Studies* 31, no. 1 (1998).

3. Dessalegn, "Littering the Landscape," 205.

4. Sebsebe Demissew, "A Study of the Vegetation and Floristic Composition of Southern Wallo, Ethiopia," in *African Savannas: Global Narratives and Local Knowledge of Environmental Change*, ed. Thomas J. Bassett and Donald Crummey (Oxford: James Currey, 2003), 159–82.

5. Belay, "Indigenous Soil Knowledge and Fertility Management."

6. Belay, "Potentials and Limitations." Belay expanded his discussion of these issues in a later article, which drew heavily on GIS technology. Belay, "Combining Land Capability Evaluation."

7. Belay, "Land Cover/Land-Use Changes."

8. Ibid., 10, 14–15.

9. Bahru, "Forests and Forest Management."

10. Crummey, "Deforestation in Wallo."

11. Information in the interviews was abstracted into a database, originally a DOS-based program called Notebook II, now in MS Access.

12. See Crummey, *Land and Society in the Christian Kingdom of Ethiopia* (again, with the assistance of Merera Ejjeta).

13. This account follows closely the account in Bassett and Crummey, *African Savannas*, 102–9.

14. Markos offers only general views gleaned from informants to the effect that there was an "acute shortage of forest resources." Markos, "Demographic Responses," 115–20.

15. One elder mentioned tanning as a source of subsidiary income.

16. The actual percent was higher because "no responses" have been included in the overall total. Markos Ezra reports that for the census period 1984–94, 48 percent of household heads report someone in the household dying, the largest number occurring during the 1984–85 famine. Markos, "Demographic Responses," 310.

17. The discussion here modifies conclusions first forwarded in Donald Crummey, "Family and Mobility in Wallo: Tahuladare Warada, 1941 to the Present," in *Ethiopian Studies at the End of the Second Millennium: Proceedings of the XIVth International Conference of Ethiopian Studies, November 6–11, Addis Ababa*, 3 vols., ed. Baye Yiman et al. (Addis Ababa: Institute of Ethiopian Studies, 2002), 1:344–54.

18. See here Bauer, *Household and Society in Ethiopia*, 123–31. On the basis of a more comprehensive social survey of a Tigray village, Bauer indicates that 25.7 percent of marriages that did not end in death ended in divorce.

19. I have discussed some of this material in Crummey, "Family and Mobility."

20. I first approached this material in "Family and Mobility." Studies of mobility in Ethiopia have tended to concentrate on urban dwellers. See, for example, Seleshi Sisaye, "Urban Migration and the Labor Movement in Ethiopia," in *Proceedings of the Fifth International Conference on Ethiopian Studies: Session B April 15–16, 1978, Chicago, USA*, ed. Robert Hess (Chicago: University of Illinois at Chicago Circle, 1979), 679–700; various contributions to *Proceedings of the Fourth Seminar of the Department*

of History, Awasa, 8–12 July 1987 (Addis Ababa: Department of History, Addis Ababa University, 1989), especially Banti Getahun, "Shashamanne: Foundation and Early Growth up to the Italian Occupation."

21. Notable studies include Alula Pankhurst, "Social Dimensions of Famine in Ethiopia: Exchange, Migration and Integration," in *Proceedings of the Ninth International Congress of Ethiopian Studies: Moscow 26–29 August, 1986*, 6 vols. (Moscow: Nauka, 1988), 2:65–80; Dessalegn Rahmato, "Some Notes on Settlement and Resettlement in Mettekel Awraja (Gojjam Province)," in *Proceedings of the Ninth International Congress of Ethiopian Studies: Moscow 26–29 August, 1986*, 6 vols. (Moscow: Nauka Publishers, 1988), 1:116–34; Dessalegn, *Famine and Survival Strategies*; Abebe Zegeye, "Environmental Degradation, Population Movement and War," in *Ethiopia in Change: Peasantry, Nationalism and Democracy*, ed. Abebe Zegeye and Siegfried Pausewang (London: British Academic Press, 1994), 172–91; and James McCann, "Household Economy, Demography, and the 'Push' Factor in Northern Ethiopian History, 1916–35," *Review* 9, no. 3 (1986): 369–411.

22. People's Democratic Republic of Ethiopia, Office of the Population and Housing Census Commission, *Population and Housing Census 1984, Analytical Report on Gondar Region* (Addis Ababa, November 1990), 233–61; People's Democratic Republic of Ethiopia, *Analytical Report on Wello Region* (Addis Ababa, January 1991), 243–69. Markos Ezra's study privileges a crisis framework. Markos, "Demographic Responses," 66–71.

23. Mesfin, *Suffering under God's Environment*, 66–71. Markos's primary interest was in the relationship between migration and stress, although he also found that overwhelmingly informants gave domestic reasons for their moving. Only 8 percent said that they moved because of "stress," such as lack of land. Markos, "Demographic Responses," 317–23.

24. To speak of peasant associations is anachronistic for much of my informants' lives. They were already in their late forties and fifties when the association system was introduced. Nonetheless, current boundaries can readily be transferred back in time.

25. See Bahru, *A History of Modern Ethiopia*, 127–28.

26. Ibid., 137–38.

27. The draft was announced in 1981 and formally instituted in 1983. Clapham, *Transformation and Continuity*, 109.

28. Most of our information about seed was anecdotal—farms produce most of their own, and farmers experiment with new seed when they find it available. Recently a common source has been the grain provided as famine relief.

29. Markos documented how gathering and fetching both water and fuelwood has taken increasingly larger amounts of time over the ten years before his study. Markos, "Demographic Responses," 117, table 5.6.

30. The classical account at the local level is by Hoben, *Land Tenure among the Amhara*. See also Bauer, *Household and Society in Ethiopia*. For a view from the standpoint of the state and its legal system, see Joireman, *Property Rights and Political Development*.

31. For the land reform, see Clapham, *Transformation and Continuity*, 161–68; Andargachew, *The Ethiopian Revolution 1974–1987*, 97–112; and Dessalegn, *Agrarian Reform in Ethiopia*.

32. See Dessalegn Rahmato, ed., *Land Tenure and Land Policy in Ethiopia after the Derg: Proceedings of the Second Workshop of the Land Tenure Project* (Trondheim: University of Trondheim, Centre for Environment and Development, 1994), especially chap. 3 (95–132); Teferi Abate, "Land Scarcity and Landlessness in North Shewa: A Case Study from Wayu and Anget Mewgiya PA"; and Svein Ege, *The Promised Land: The Amhara Land Redistribution of 1997* (Dragvoll: Norwegian University of Science and Technology, Centre for Environment and Development, 1997).

33. Markos's calculation was that there were 0.75 hectare of cultivated land per household ("Demographic Responses," 68); whereas Mesfin reckoned this number to be 1.0 hectare (*Suffering under God's Environment*, 89). Given that not one of us was able to conduct a true measurement, we should expect some variation in findings. That it should be to this degree is another matter.

34. Bauer, *Household and Society in Ethiopia*, chap. 4.

35. Mesfin reports average holdings per farmer as 1.7 oxen, 2.7 cows and heifers, 5.9 small stock, and 0.8 equines. Mesfin, *Suffering under God's Environment*, 111, table 5.3. I have collapsed a number of his categories.

36. Dessalegn, *Famine and Survival Strategies*, 171–76.

37. Ninety-eight percent of Mesfin's informants reported shortages of or problems with grazing land. Mesfin, *Suffering under God's Environment*, 114–15.

38. Darg policy and action, in its insensitivity to local interests and needs, bears close comparison with the behavior of colonial governments in their declining days, where heavy-handed conservation measures fueled the growth of African nationalism.

39. Esther Boserup, *The Conditions of Agricultural Change: The Economics of Agrarian Change under Population Pressure* (New York: Aldine, 1965). For a somewhat different view, see Markos, "Demographic Responses," 28–29.

40. Winter-Nelson, "Rural Taxation in Ethiopia"; as cited in Crummey and Winter-Nelson, "Farmer Tree-Planting in Wallo," 99. This passage is heavily dependent on the corresponding passage in Crummey and Winter-Nelson.

41. This information may be compared to table 5.6, "Stated Uses for Trees," in Crummey and Winter-Nelson, "Farmer Tree-Planting in Wallo," 101.

42. Again, the discussion follows closely the discussion in Crummey and Winter-Nelson, "Farmer Tree-Planting in Wallo," 109–11.

43. Holt and Lawrence, *Making Ends Meet*.

44. Richard Pankhurst provides a useful account in *Economic History of Ethiopia 1800–1935* (Addis Ababa: Haile Selassie I University Press, 1968), 245–47.

45. The Maugini photographs establish the presence in Dessie, for which see also Lusana, "L'Uollo Orientale."

BIBLIOGRAPHY

Abdul Mejid. "The Political Economy of the Famine in Ethiopia." In *Rehab: Drought and Famine in Ethiopia*, edited by Abdul Meijid Hussein, 9–51. London: International African Institute, 1976.

———, ed. *Drought and Famine in Ethiopia*. London: International African Institute, 1976.

Abdussamad H. Ahmad. "Peasant Conditions in Gojjam during the Great Famine 1888–1921." *Journal of Ethiopian Studies* 20 (November 1987): 10–18.

Abdussamad H. Ahmad and Richard Pankhurst, eds. *Adwa Victory Centenary Conference, 26 February–2 March 1996*. Addis Ababa: Institute of Ethiopian Studies, Addis Ababa University, 1998.

Abebe Zegeye. "Environmental Degradation, Population Movement and War." In *Ethiopia in Change: Peasantry, Nationalism and Democracy*, edited by Abebe Zegeye and Siegfried Pausewang, 172–91. London: British Academic Press, 1994.

Adhana Haile Adhana. "History of Selected Famines in Peasant Societies in Tigray and Wallo, Ethiopia, 1941–1974." PhD diss., Addis Ababa University, 1996.

Afawarq Gabra Iyasus. *Dagmawi Ase Menilek*. Rome: n.p., 1927.

Ahmed, Hussein. *Islam in Nineteenth-Century Wallo, Ethiopia: Revival and Reaction*. Leiden: Brill, 2001.

Alaqa Lamma. "Tarik Yasme Takla Giyorgis[enna] Yatse Yohannes." Mss. Éthiopien d'Abbadie 259, folio 36. Bibliothèque Nationale, Paris, n.d.

Amsalu Aklilu. *Amaregna-Englizegna Mazgaba Qalat/Amharic-English Dictionary*. Addis Ababa: Kuraz Asātāmi Drjt, 1979 EC.

Andargachew Tiruneh. *The Ethiopian Revolution 1974–1987: A Transformation from an Aristocratic to a Totalitarian Autocracy*. Cambridge: Cambridge University Press, 1993.

Appuhn, Karl. "Ecologies of Beef: Eighteenth-Century Epizootics and the Environmental History of Early Modern Europe." *Environmental History* 15, no. 2 (2010): 268–87.

Arén, Gustav. *Evangelical Pioneers in Ethiopia: Origins of the Evangelical Church Mekane Yesus*. Stockholm: EFS Förlaget; 1978.

Aspen, Harald. *Competition and Co-operation: North Ethiopian Peasant Households and Their Resource Base*. Working Paper on Ethiopian Development no. 7. Trondheim: Trondheim Universitetet Centre for Environment and Development, 1993.
Bahru Zewde. "Forests and Forest Management in Wallo in Historical Perspective." *Journal of Ethiopian Studies* 31, no. 1 (1998): 87–121.
———. "A Historical Outline of Famine in Ethiopia." In *Rehab: Drought and Famine in Ethiopia*, edited by Abdul Meijid Hussein, 52–58. London: International African Institute, 1976.
———. *A History of Modern Ethiopia 1855–1991*. 2nd ed. Oxford: James Currey, 2001.
Bairu Tafla, ed. *Asma Giyorgis and His Work: History of the Galla and the Kingdom of Šawa*. Stuttgart: Frank Steiner, 1987.
———, ed. *A Chronicle of Emperor Yohannes IV (1872–89)*. Wiesbaden: Franz Steiner, 1977.
———, ed. *Ethiopian Records of the Menilek Era: Selected Amharic Documents from the Nachlass of Alfred Ilg 1884–1900*. Wiesbaden: Harrassowitz, 2000.
Balsvik, Randi R. *Haile Sellassie's Students: The Intellectual and Social Background to Revolution, 1952–1977*. East Lansing: African Studies Center, Michigan State University, 1985.
Banti Getahun. "Shashamanne: Foundation and Early Growth up to the Italian Occupation." In *Proceedings of the Fourth Seminar of the Department of History, Awasa, 8–12 July 1987*, 10–27. Addis Ababa: Department of History, Addis Ababa University, 1989.
Bassett, Thomas J., and Donald Crummey, eds. *African Savannas: Global Narratives and Local Knowledge of Environmental Change*. Oxford: James Currey, 2003.
———. "Contested Images, Contested Realities: Environment and Society in African Savannas." In *African Savannas: Global Narratives and Local Knowledge of Environmental Change*, edited by Thomas J. Bassett and Donald Crummey, 1–30. Oxford: James Currey, 2003.
Bauer, Dan Franz. *Household and Society in Ethiopia*. East Lansing: African Studies Center, Michigan State University, 1977.
Baulch, Bob. "Entitlements and the Wollo Famine of 1982–1985." *Disasters* 11, no. 3 (September 1987): 195–204.
Belay Tegene. "Combining Land Capability Evaluation, Geographic Information Systems, and Indigenous Technologies for Soil Conservation in Northern Ethiopia." *East African Social Science Research Review* 19, no. 2 (June 2003): 23–53.
———. "Indigenous Soil Knowledge and Fertility Management Practices of the South Wallo Highlands." *Journal of Ethiopian Studies* 31, no. 1 (1998): 123–58.
———. "Land Cover/Land-use Changes in the Derekolli Catchment of the South Welo Zone of Amhara Region, Ethiopia." *Eastern Africa Social Science Research Review* 18, no. 1 (January 2002): 1–20.
———. "Potentials and Limitations of an Indigenous Structural Soil Conservation Technology of Welo, Ethiopia." *Eastern Africa Social Science Research Review* 14, no. 1 (January 1998): 1–18.
Abenet Belete, John L. Dillon, and Frank M. Anderson. "Development of Agriculture in Ethiopia since the 1975 Land Reform." *Agricultural Economics* 6 (1991): 159–75.

Shewandagn Belete, Mehari Gebre-Mehin, Bantirou Hailemariam, Mario Maffi, Bo Vahlquist, and Zewdie Wolde-Gebriel. "Study of Shelter Population in the Wollo Region." *Journal of Tropical Pediatrics and Environmental Child Health* 23, no. 1 (February 1977). Monograph 48, "Famine in Ethiopia," edited by Bo Vahlquist, 15–22.

Bent, J. Theodore. *The Sacred City of the Ethiopians Being a Record of Travel and Research in Abyssinia in 1893*. London, 1896.

Berhanu Abegaz. "Persistent Stasis in a Tributary Mode of Production: The Peasant Economy of Ethiopia." *Journal of Agrarian Change* 5, no. 3 (July 2005): 299–333.

Boerma, Pauline A. "Assessing Forest Cover Change in Eritrea—A Historical Perspective." *Mountain Research and Development* 25, no. 1 (2006): 41–47.

———. "Myths, Memories, and Metaphors: Recollecting Landscape Changes in the Eritrean Highlands." *Journal of Eastern African Studies* 6, no. 2 (2012): 246–69.

———. "Seeing the Wood for the Trees: Deforestation in the Central Highlands of Eritrea since 1890." PhD diss., University of Oxford, 1999.

Boserup, Esther. *The Conditions of Agricultural Change: The Economics of Agrarian Change under Population Pressure*. New York: Aldine, 1965.

Brielli, D. "Ricordi storici dei Uollo (con note di C. Conti Rossini)." In *Studi Etiopici*, edited by C. Conti Rossini, 78–110. Rome: Istituto per l'Oriente, 1945.

Brown, Lester R., and Edward C. Wolfe. *Reversing Africa's Decline*. Worldwatch Paper 65. Washington, DC: Worldwatch Institute, 1985.

Bulatovich, Alexander. *Ethiopia through Russian Eyes: Country in Transition 1896–1898*. Translated and edited by Richard Seltzer. Lawrenceville, NJ: Red Sea Press, 2000.

Carlson, Andrew J., and Dennis G. Carlson. *Health, Wealth, and Family in Rural Ethiopia: Kossoye, North Gondar Region, 1963–2007*. Addis Ababa: Addis Ababa University Press, 2008.

Caulk, R. A. "Armies as Predators: Soldiers and Peasants in Ethiopia c. 1850–1935." *International Journal of African Historical Studies* 11, no. 3 (1978): 457–93.

Caulk, Richard. *"Between the Jaws of Hyenas": A Diplomatic History of Ethiopia 1876–1896*. Edited by Bahru Zewde. Wiesbaden: Harrassowitz, 2002.

Caulk, Richard. "Religion and the State in Nineteenth Century Ethiopia." *Journal of Ethiopian Studies* 10, no. 1 (1972): 23–41.

Cerroti, O. "Il secondo Viaggio in Abissinia del Mashcov (1891–1892)." *Bolletino della Società GeograficaItaliana*, series 3, vol. 7 (1894): 841–86.

Clapham, Christopher. *Haile-Selassie's Government*. London: Longmans, Green, 1969.

———. *Transformation and Continuity in Revolutionary Ethiopia*. Cambridge: Cambridge University Press, 1988.

Clay, Jason W., and Bonnie K. Holcomb. *Politics and the Ethiopian Famine 1984–1985*. Cambridge: Cultural Survival, 1986.

Cliffe, Lionel. "The Impact of War and the Response to It in Different Agrarian Systems in Eritrea." *Development and Change* 20, no. 3 (1989): 373–400.

Cohen, John M. "Ethiopia after Haile Selassie: The Government Land Factor." *African Affairs* 72, no. 289 (1973): 365–82.

Constable, Michael, and Deryke Belshaw. "The Ethiopian Highlands Reclamation Study: Major Findings and Recommendations." In *Towards a Food and Nutrition*

Strategy for Ethiopia: The Proceedings of the National Workshop on Food Strategies for Ethiopia, Held at Alemaya University of Agriculture, 8–12 December 1986, 142–79. Addis Ababa: People's Democratic Republic of Ethiopia, Office of the National Committee for Central Planning, 1989.

Conway, D. "Some Aspects of Climate Variability in the North East Ethiopian Highlands—Wollo and Tigray." *SINET: An Ethiopian Journal of Science* 23, no. 2 (2000): 139–61.

Conway, D., and E. Lisa F. Schipper. "Adaptation to Climate Change in Africa: Challenges and Opportunities Identified from Ethiopia." *Global Environmental Change* 21 (2011): 227–37.

Corigliano, M. "Regioni Poco note dell'Africa O. I. La residenza del lago Haik." *Illustrazione Coloniale* 19, no. 3 (March 1941): 33–37.

Crawford, Eric W., and Gregory C. Lassiter. "Constraints on Oxen Cultivation in the Sahel: Comment." *American Journal of Agricultural Economics* 67, no. 3 (1985): 188–96.

Crummey, Donald. "Church and Nation: The Ethiopian Orthodox *Täwahedo* Church (from the Thirteenth to the Twentieth Century)." In *Cambridge History of Christianity*, vol. 5, *Eastern Christianity*, edited by Michael Angold, 457–87. Cambridge: Cambridge University Press, 2006.

———. "Deforestation in Wallo: Process or Illusion?" *Journal of Ethiopian Studies* 36, no. 1 (1998): 1–41.

———. "Exploring Landscape Change in Ethiopia: Evidence from Imaging and Its Interpretation." In *Research in Ethiopian Studies: Selected Papers of the 16th International Conference of Ethiopian Studies, Trondheim July 2007*, edited by Harald Aspen, Birhanu Teferra, Shiferaw Bekele, and Svein Ege, 93–105. Wiesbaden: Harrassowitz, 2010.

———. "Family and Mobility in Wallo: Tahuladare Warada, 1941 to the Present." In *Ethiopian Studies at the End of the Second Millennium: Proceedings of the XIVth International Conference of Ethiopian Studies, November 6–11, Addis Ababa*, vol. 1, edited by Baye Yiman et al., 344–54. Addis Ababa: Institute of Ethiopian Studies, 2002.

———. "Imperial Legitimacy and the Creation of a Neo-Solomonic Ideology in 19th Century Ethiopia." *Cahiers d'Études africaines* 109, no. 28 (1988): 13–43.

———. *Land and Society in the Christian Kingdom of Ethiopia: From the Thirteenth to the Twentieth Century*. Urbana: University of Illinois Press, 2000.

Crummey, Donald, and Alex Winter-Nelson. "Farmer Tree-Planting in Wallo, Ethiopia, 1937–1997." In *African Savannas: Global Narratives and Local Knowledge of Environmental Change*, edited by Thomas J. Bassett and Donald Crummey, 91–120. Oxford: James Currey, 2003.

Cutler, Peter. "Famine Forecasting: Prices and Peasant Behavior in Northern Ethiopia." *Disasters* 6, no. 1 (1984): 48–56.

Dawit Wolde Giorgis. *Red Tears: War Famine and Revolution in Ethiopia*. Trenton, NJ: Red Sea Press, 1989.

Alemneh Dejene. *Environment, Famine, and Politics in Ethiopia: A View from the Village*. Boulder, CO: Lynne Rienner, 1990.

———. *Integrated Natural Resources Management to Enhance Food Security: The Case for Community-Based Approaches in Ethiopia*. Rome: UN FAO, 2003.

Delgado, Christopher L., and John McIntire. "Constraints on Oxen Cultivation in the Sahel." *American Journal of Agricultural Economics* 64, no. 2 (1982): 188–96.

———. "Constraints on Oxen Cultivation in the Sahel: Reply." *American Journal of Agricultural Economics* 67, no. 3 (1985): 686–87.

de Waal, Alex. *Famine Crimes: Politics and the Disaster Relief Industry in Africa*. London/Oxford/Bloomington: African Rights and the International African Institute/James Currey/Indiana University Press, 1997.

Derseh Endale. *Rural Markets, Food-Grain Prices and Famines: A Study on Selected Regions in Ethiopia*. Working Paper no. 118. Washington, DC: United Nations University World Institute for Development Economics Research, 1995.

Dessalegn Rahmato. *Agrarian Reform in Ethiopia*. Trenton, NJ: Red Sea Press, 1985.

———. "Agrarian Reform in Ethiopia: A Brief Assessment." In *Proceedings of the Seventh International Conference of Ethiopian Studies*, edited by Sven Rubenson, 587–95. Addis Ababa: Institute of Ethiopian Studies, 1984.

———. *Famine and Survival Strategies: A Case Study from Northeast Ethiopia*. Uppsala: Nordiska Afrikainstitutet, 1991.

———, ed. *Land Tenure and Land Policy in Ethiopia after the Derg: Proceedings of the Second Workshop of the Land Tenure Project*. Trondheim: University of Trondheim, Centre for Environment and Development, 1994.

———. "Littering the Landscape: Environmental Policy in Northeast Ethiopia." In *African Savannas: Global Narratives and Local Knowledge of Environmental Change*, edited by Thomas J. Bassett and Donald Crummey, 205–24. Oxford: James Currey, 2003.

———. "Some Notes on Settlement and Resettlement in Mettekel Awraja (Gojjam Province)." In *Proceedings of the Ninth International Congress of Ethiopian Studies: Moscow 26–29 August, 1986*, 6 vols., 116–34. Moscow: Nauka, 1988.

Devereux, Stephen. "Entitlements, Availability and Famine: A Revisionist View of Wollo, 1972–74." *Food Policy* (August 1988): 270–82.

Ege, Svein. *The Promised Land: The Amhara Land Redistribution of 1997*. Dragvoll: Norwegian University of Science and Technology, Centre for Environment and Development, 1997.

Erlich, Haggai. *Ethiopia and Eritrea during the Scramble for Africa: A Political Biography of Ras Alula, 1875–1897*. East Lansing, MI: African Studies Center, Michigan State University, 1982.

Fairhead, James, and Melissa Leach. *Misreading the African Landscape: Society and Ecology in a Forest-Savanna Mosaic*. Cambridge: Cambridge University Press, 1996.

Fasil Nahum. *Constitution for a Nation of Nations: The Ethiopian Prospect*. Lawrenceville, NJ: Red Sea Press, 1997.

Fekadu Begna. "A Tentative History of Wallo, 1855–1908." Bachelor's thesis, Addis Ababa University, 1972.

Flad, Johann Martin. *60 Jahre in der Mission unter den Falaschas in Abessinien: Selbstbiographie des Missionars Johann Martin Flad: Mit Einleitung und Schlusswort von seinem Sohn Pastor Friedrich Flad*. Giessen and Basel, 1922.

Ford, John. *The Role of Trypanosomiases in African Ecology: A Study of the Tsetse Fly Problem*. Oxford: Clarendon Press, 1971.

Gabra Sellasie. *Chronique du règne du Ménélik II, Roi des Rois d'Éthiopie, d'après un Manuscrit Original.* Edited and translated by Casimir Mondon-Vidailhet. 2 vols. Paris, 1930, 1932.

———. *Tarik zaman za Dagmawi Menilek negusa nagast za-Ityopya.* Addis Ababa, 1959.

Zewde Gabre-Sellassie. *Yohannes IV of Ethiopia: A Political Biography.* Oxford: Clarendon Press, 1975.

Gascon, Alain. "La forêt perdue d'Éthiopie, un mythe contemporain." In *Plantes et paysages d'Afrique: Une histoire à explorer,* edited by Monique Chastanet, 390–98. Paris: Editions Karthala, Centre de recherches africaines, 1998.

———. *Sure les hautes terres comme au ciel: Identités et territoires en Ethiopie.* Paris: Publications de la Sorbonne, 2006.

Gebru Tareke. *The Ethiopian Revolution: War in the Horn of Africa.* New Haven, CT: Yale University Press, 2009.

———. *Power and Protest: Peasant Revolts in the Twentieth Century.* Cambridge: Cambridge University Press, 1991.

Gete Zeleke and Hans Hurni. "Implications of Land Use and Land Cover Dynamics for Mountain Resource Degradation in the Northwestern Ethiopian Highlands." *Mountain Research and Development* 22, no. 2 (May 2001): 184–91.

Gilkes, Patrick. *The Dying Lion: Feudalism and Modernization in Ethiopia.* New York: St. Martin's Press, 1975.

Gill, Peter. *A Year in the Death of Africa: Politics, Bureaucracy and the Famine.* London: Paladin, 1986.

Girmay Kassa Welela. "GIS Based Analysis of Land Use/Land Cover, Land Degradation and Population Changes (A Study of Boru-Metero Area of South Wello, Amhara Region)." Master's thesis, Addis Ababa University, 2003.

Glantz, M. H., ed. *Drought Follows the Plow.* Cambridge: Cambridge University Press, 1994.

———, ed. *The Politics of Natural Disaster: The Case of Sahel Drought.* New York: Praeger, 1976.

Glantz, Michael H., and Workineh Degefu. "Drought Issues for the 1990s." In *Climate Change: Science, Impacts and Policy: Proceedings of the Second World Climate Conference,* edited by J. Jäger and J. L. Ferguson, 253–63. Cambridge: Cambridge University Press, 1991.

Grove, Richard H. *Green Imperialism: Colonial Expansion, Tropical Island Edens and the Origins of Environmentalism, 1600–1860.* Cambridge: Cambridge University Press, 1995.

Grunder, M. "Soil Conservation Research in Ethiopia." *Mountain Research and Development* 8, nos. 2/3 (May/August 1988): 145–51.

Guidi, I. *Supplemento al Vocabolario Amarico-Italiano Compilata con il Concorso di Francesco Gallina ed Enrico Cerulli.* Rome: Istituto per l'Oriente, 1940.

Hahn, E. *Die Entstehung der Pflugkultur (unsres Ackerbaus).* Heidelberg, 1909.

Haile Mariam Larebo. *The Building of an Empire: Italian Land Policy and Practice in Ethiopia 1935–1941.* Oxford: Clarendon Press, 1994.

Halliday, Fred, and Maxine Molyneux. *The Ethiopian Revolution.* London: Verso, 1981.

Hansson, Göte. *The Ethiopian Economy 1974–94: Ethiopia Tikdem and After*. New York: Routledge, 1995.

Harbeson, John. *The Ethiopian Transformation: The Quest for the Post-imperial State*. Boulder, CO: Westview Press, 1987.

Harlan, Jack R. *The Living Fields: Our Agricultural Heritage*. Cambridge: Cambridge University Press, 1995.

Hassan, Fekri A. "Historical Nile Floods and Their Implications for Climatic Change." *Science* 212, no. 4499 (June 5, 1981): 1142–45.

Haudricourt, André G., and Mariel Jean-Brunhes Delamarre. *L'Homme et la charrue à travers le monde*. Paris, 1955.

Helldén, U. "Desertification—Time for an Assessment?" *Ambio* 20, no. 8 (1991): 372–83.

Henze, Paul. *The Primacy of Economics for the Future of the Horn of Africa*. Santa Monica, CA: Rand, 1992.

Hoben, A. *Land Tenure among the Amhara of Ethiopia: The Dynamics of Cognatic Descent*. Chicago: University of Chicago Press, 1973.

———. "Paradigms and Politics: The Cultural Construction of Environmental Policy in Ethiopia." *World Development* 23, no. 6 (1995): 1007–21.

Holt, Julius, and Mark Lawrence. *Making Ends Meet: A Survey of the Food Economy of the Ethiopian North-East Highlands*. London: Save the Children UK, 1993.

Holt, Julius, and John Seaman. "Markets and Famines in the Third World." *Disasters* 4, no. 3 (1980): 283–97.

———. "The Scope of the Drought." In *Rehab: Drought and Famine in Ethiopia*, edited by Abdul Meijid Hussein, 1–8. London: International African Institute, 1976.

Holt, Julius, John Seaman, and John Rivers. "Famine Revisited." *Nature* 255 (May 15, 1975): 180–81.

Huffnagel, H. P. *Agriculture in Ethiopia*. Rome: FAO, 1961.

Hurni, Hans. *Climate and Dynamics of Altitudinal Belts from the Last Cold Period to the Present Day*. Vol. 2 of *Simien Mountains: Ethiopia*. Bern: Geographishches Institut der Universität Bern, 1982.

———. "Degradation and Conservation of the Resources in the Ethiopian Highlands." *Mountain Research and Development* 8, nos. 2/3 (May/August 1988): 123–30.

———. "Harvesting Areas and Cultivation Belts in and around the Simien Mountains National Park—Ethiopia." n.d.

Istituto Agronomico per l'Oltremare. *Armando Maugini nel centenario della nascita*. Florence, 1989.

Iyob, Ruth. *The Eritrean Struggle for Independence: Dominance, Resistance, Nationalism 1941–1993*. Cambridge: Cambridge University Press, 1995.

James, Wendy, Donald L. Donham, Eisei Kurimoto, and Alessandro Triulzi. *Remapping Ethiopia: Socialism and After*. Oxford: Ohio University Press, 2002.

Jansson, Kurt, Michael Harris, and Angela Penrose. *The Ethiopian Famine*. London: Zed Books, 1987.

Joireman, Sandra F. *Property Rights and Political Development in Ethiopia and Eritrea*. Oxford: James Currey, 2000.

Kaplan, Robert D. *Surrender or Starve: The Wars behind the Famine*. Boulder, CO: Westview Press, 1988.

Kebede Tadesse. *Trees of Ethiopia: A Photographic Guide and Description*. Addis Ababa: Washera, 2012.

Kebrom Tekle and Lars Hedlund. "Land Cover Changes between 1958 and 1986 in Kalu District, Southern Wello, Ethiopia." *Mountain Research and Development* 20, no. 1 (February 2000): 42–51.

Keller, Conrad. *Alfred Ilg: Sein Leben and sein Wirken als schweizerischer Kulturbote in Abessinien*. Frauenfeld and Leipzig: Huber, 1918.

Keller, Edmond J. "Drought, War and the Politics of Famine in Ethiopia and Eritrea." *Journal of Modern African Studies* 30, no. 4 (1992): 609–24.

———. *Revolutionary Ethiopia: From Empire to People's Republic*. Bloomington: Indiana University Press, 1990.

Gaim Kibreab. *People on the Edge in the Horn*. Oxford: James Currey, 1996.

Fassil G. Kiros. *Enough with Famines in Ethiopia. A Clarion Call.* Addis Ababa, 2005.

Kolmodin, Johannes. *Traditions de Tsazzega et Hazzega*. 2 vols. Upsala, Appelberg, 1915–16.

Kreike, E. *Re-creating Eden: Land Use, Environment and Society in Southern Angola and Northern Namibia*. Portsmouth, NH: Heinemann, 2004.

Kumar, G. *Ethiopian Famines 1973–1985: A Case-Study*. Helsinki: World Institute for Development Economics Research, 1987.

Mangestu Lamma. *Mashafa Tezeta ZaAlaqa Lamma Haylu Walda Tarik*. Addis Ababa, 1959 Eth. Cal.

Lauribar, Paul de. *Douze Ans en Abyssinie (Souvenirs d'un Officier)*. Paris, 1898.

Lefort, René. *Ethiopia, an Heretical Revolution?* London: Zed, 1983.

Legum, Colin. *Ethiopia: The Fall of Haile Selassie's Empire*. New York: Africana, 1975.

Li, Lillian M. "Famine and Famine Relief: Viewing Africa in the 1980s from China in the 1920s." In *Drought and Hunger in Africa: Denying Famine a Future*, edited by Michael Glantz, 415–34. Cambridge: Cambridge University Press, 1987.

Lockot, Hans Wilhelm. *The Mission: The Life, Reign and Character of Haile Sellassie I.* London: Hurst and Company, 1989.

Lonsdale, John. "The European Scramble and Conquest in African History." In *The Cambridge History of Africa*, vol. 6, *1870–1905*, edited by Roland Oliver and G. N. Sanderson, 689–90. Cambridge: Cambridge University Press, 1985.

Lusana, Alessandro. "L'Uollo Orientale." *Gli Annali dell'Africa Italiana* 2, no. 2 (1939): 477–515.

Lusana, Alessandro. "L'Itinerari dell'A. O. I. La regione di Dessie." *L'Azione Coloniale* 10, no. 17 (April 25, 1940): 4.

Marcovitch, Harvey. *Black's Veterinary Dictionary*. 41st ed. London: A&C Black, 2005.

Marcus, Harold G. *Haile Sellassie I: The Formative Years, 1892–1936*. Berkeley and Los Angeles: University of California Press, 1987.

———. *The Life and Times of Menelik II: Ethiopia, 1844–1913*. Lawrenceville, NJ: Red Sea Press, 1995.

Markakis, John. *Ethiopia: Anatomy of a Traditional Polity*. Oxford: Clarendon Press, 1974.

Markakis, John, and Nega Ayele. *Class and Revolution in Ethiopia*. Nottingham: Spokesman, 1978.

Markos Ezra. "Demographic Responses to Ecological Degradation and Food Insecurity: Drought Prone Areas in Northern Ethiopia." PhD diss., Royal University of Groningen, 1997.

Markos Ezra and Kassahun Berhanu. *A Review of the Community Forestry Programme and an Evaluation of Its Achievements: A Socio-Economic Survey*. Addis Ababa: Swedish International Development Agency, 1988.

Marks, Shula. "Southern and Central Africa, 1886–1910." In *Cambridge History of Africa*, vol. 6, *From 1870 to 1905*, edited by Roland Oliver and G. N. Sanderson, 422–92. Cambridge: Cambridge University Press, 1985.

Martini, Ferdinando. *Nell'Affrica Italiana*. Milan: Fratelli Treves, 1925.

Mason, J. B., R. W. Hay, J. Holt, J. Seaman, and M. R. Bowden. "Nutritional Lessons from the Ethiopian Drought." *Nature* 248 (April 19, 1974): 646–50.

Maugini, Elena, and Luca Fabbri. "Storia di una vita: note biografiche di e su Armando Maugini." Mimeograph in the Biblioteca della Facoltà di Agraria, Firenze, Fondo Maugini, 1948.

McCann, James. "Ethiopia." In *Drought Follows the Plow*, edited by M. H. Glantz, 103–15. Cambridge: Cambridge University Press, 1994.

———. *From Poverty to Famine in Northeast Ethiopia: A Rural History, 1900–1935*. Philadelphia: University of Pennsylvania Press, 1987.

———. *Green Land, Brown Land, Black Land: An Environmental History of Africa, 1800–1990*. Portsmouth, NH: Heinemann, 1999.

———. "Household Economy, Demography, and the 'Push' Factor in Northern Ethiopian History, 1916–35." *Review* 9, no. 3 (1986): 369–411.

———. *People of the Plow: An Agricultural History of Ethiopia, 1800–1990*. Madison: University of Wisconsin Press, 1995.

———. "The Plow and the Forest: Narratives of Deforestation in Ethiopia, 1840–1992." *Environmental History* 2, no. 2 (1997): 138–59.

———. "The Social Impact of Drought in Ethiopia: Oxen, Households, and Some Implications for Rehabilitation." In *Drought and Hunger in Africa: Denying Famine a Future*, edited by Michael Glantz, 245–67. Cambridge: Cambridge University Press, 1987.

McClellan, Charles W. *State Transformation and National Integration: Gedeo and the Ethiopian Empire, 1895–1935*. East Lansing: African Studies Center, Michigan State University, 1988.

Kidane Mengisteab. *Ethiopia: Failure of Land Reform and Agricultural Crisis*. New York: Greenwood Press, 1990.

Mehari Gebre-Medhin and Bo Vahlquist. "Famine in Ethiopia—The Period 1973–75." *Nutrition Reviews* 35, no. 8 (August 1977): 194–202.

Mesfin Wolde Mariam. *An Introductory Geography of Ethiopia*. Addis Ababa, 1972.

———. *Rural Vulnerability to Famine in Ethiopia 1958–1977*. New Delhi: Vikas Publishing House, 1984.

———. *Suffering under God's Environment: A Vertical Study of the Predicament of Peasants in North-Central Ethiopia*. Berne: African Mountains Association and Geographica Bernensia, 1991.

Meze-Hausken, Elisabeth. "Contrasting Climate Variability and Meteorological Drought with Perceived Drought and Climate Change in Northern Ethiopia." *Climate Research* 27 (2004): 19–31.

Mortimer, Michael. *Adapting to Drought: Farmers, Famines and Desertification in West Africai*. Cambridge: Cambridge University Press, 1987.

Nolan, Liam. *The Forgotten Famine*. Dublin: Mercier Press, 1974.

Nyssen, J., A. Frankl, R. N. Munro, P. Billi, and M. Haile. "Digital Photographic Archives for Environmental and Historical Studies: An Example from Ethiopia." *Scottish Geographical Journal* 126, no. 3 (2010): 185–207.

Nyssen, J., M. Haile, J. Naudts, R. N. Munro, J. Poesen, J. Moeyersons, A. Frankl, J. Deckers, and R. Pankhurst. "Desertification? Northern Ethiopia Re-photographed after 140 Years." *Science of the Total Environment* 407 (2009): 2749–55.

Ottaway, Marina, and David Ottaway. *Ethiopia: Empire in Revolution*. New York: Africana, 1978.

Pankhurst, Alula. *Resettlement and Famine in Ethiopia: The Villagers' Experience*. Manchester: Manchester University Press, 1992.

———. "Social Dimensions of Famine in Ethiopia: Exchange, Migration and Integration." In *Proceedings of the Ninth International Congress of Ethiopian Studies: Moscow 26–29 August, 1986*, 2: 65–80. Moscow: Nauka, 1988.

Pankhurst, Alula, and François Piguet, eds. *People, Space and the State. Migration, Resettlement and Discplacement in Ethiopia*. Addis Ababa: Ethiopian Society of Sociologists, Social Workers and Anthropologists, 2004.

Pankhurst, Richard. *Economic History of Ethiopia 1800–1935*. Addis Ababa: Haile Selassie I University Press, 1968.

———. "The Genesis of Photography in Ethiopia and the Horn of Africa." *British Journal of Photography* (October 8, 15, 22, and 29, 1976), 878–82, 910–13, 933–35, and 952–53.

———. "The Great Ethiopian Famine of 1888–1892: A New Assessment." *Journal of the History of Medicine and Allied Sciences* 21 (1966): 95–124 and 271–94.

———. "The Hedar Bäšeta of 1918." *Journal of Ethiopian Studies* 13, no. 2 (1975): 103–31.

———. "A Historical Note on Influenza in Ethiopia." *Medical History* 21 (1977): 195–200.

———. "The History and Traditional Treatment of Smallpox in Ethiopia." *Medical History* 9 (1965): 343–55.

———. "The History of Cholera in Ethiopia." *Medical History* 12 (1968): 262–69.

———. *The History of Famine and Epidemics in Ethiopia prior to the Twentieth Century*. Addis Ababa: Relief and Rehabilitation Commission, 1986.

———. *An Introduction to the Medical History of Ethiopia*. Trenton, NJ: Red Sea Press, 1990.

———. "Some Notes for the History of Typhus in Ethiopia." *Medical History* 20 (1976): 384–93.

Pankhurst, Richard, and Denis Gerard. *Ethiopia Photographed: Historic Photographs of the Country and Its People Taken between 1867 and 1935*. London: Kegan Paul, 1996.

Pankhurst, Richard, and Douglas H. Johnson. "The Great Drought and Famine of 1888–92 in Northeast Africa." In *The Ecology of Survival: Case Studies from Northeast African History*, edited by David M. Anderson and Douglas H. Johnson, 47–70. Boulder, CO: Westview Press, 1988.

People's Democratic Republic of Ethiopia. *Analytical Report on Wello Region*. Addis Ababa, January 1991.

People's Democratic Republic of Ethiopia, Office of the Population and Housing Census Commission. *Population and Housing Census 1984, Analytical Report on Gondar Region*. Addis Ababa, 1990.

Perham, M. *The Government of Ethiopia*. London: Faber and Faber, 1948.

Pingali, Prabhu, Yves Bigot, and Hans P. Binswanger. *Agricultural Mechanization and the Evolution of Farming Systems in Sub-Saharan Africa*. Baltimore: Johns Hopkins University Press for the World Bank, 1987.

Potter, C. W. "A History of Influenza." *Journal of Applied Microbiology* 91, no. 4 (2001): 572–79.

Prouty Rosenfeld, Chris. *A Chronology of Emperor Menilek II of Ethiopia*. East Lansing: African Studies Center, Michigan State University, 1976.

Rasmusson, Eugene M. "Global Climate Change and Variability: Effects on Drought and Desertification in Africa." In *Drought and Hunger in Africa: Denying Famine a Future*, edited by Michael H. Glantz, 3–22. Cambridge: Cambridge University Press, 1987.

Reid, Robin S., Russell L. Kruska, Nyawira Muthui, Andualem Taye, Sara Wotton, Cathleen J. Wilson, and Wondyalew Mulatu. "Land-Use and Land-Cover Dynamics in Response to Changes in Climatic, Biological and Socio-Political Forces: The Case of Southwestern Ethiopia." *Landscape Ecology* 15 (2000): 339–55.

Ritler, Alfons. *Forests, Land Use and Landscape in the Central and Northern Ethiopian Highlands, 1865 to 1930*. Publication A 19 in the African Studies Series of Geographica Bernensia. Bern: University of Berne, Institute of Geography, 2003.

———. *Land Use, Forests and the Landscape of Ethiopia, 1699–1865: An Enquiry into the Historical Geography of Central-Northern Ethiopia*. Soil Conservation Research Programme Ethiopia, Research Report no. 38. Bern: University of Bern, 1997.

———. "Wald, Landnutzung und Landschaft im Zentralen und Nördlichen Äthiopien 1865–1930." Centre for Development and Environment, Bern, 2001.

Rivers, J. P. W., J. F. J. Holt, J. A. Seaman, and M. R. Bowden. "Lessons for Epidemiology from the Ethiopian Famines." *Annales de la Société belge de médecine tropicale* 58, nos. 4–5 (1976): 345–60.

Rohde, Richard. "Nature, Cattle Thieves and Various Other Midnight Robbers: Images of People, Place and Landscape in Damaraland, Namibia." PhD diss., Edinburgh University, 1997.

Rubenson, Sven. "Adwa, 1896: The Resounding Protest." In *Protest and Power in Black Africa*, edited by Robert Rotberg and Ali Mazrui, 113–42. New York: Oxford University Press, 1970.

———. *The Survival of Ethiopian Independence*. London: Heinemann Educational Books, 1976.

Ruthenberg, Hans. *Farming Systems in the Tropics*. 2nd ed. Oxford: Clarendon Press, 1976.

Salih, M. A. Mohamed, and John Markakis. *Ethnicity and the State in Eastern Africa*. Uppsala: Nordiska Afrikainstitutet, 1998.

Salimbeni, Augusto. *Crispi e Menelich nel Diario inedito del conte Augusto Salimbeni: Con Prefazione, Introduzione, Note e Appendici*. Edited by Carlo Zaghi. Turin: Istituto per la storia del Risorgimento italiano, 1956.

———. *La Conquista dell'Africa: Studi e Ricerche*. Edited by Carlo Zaghi. 2 vols. Napoli: Istituto Universitario Orientale, 1984.

Savadogo, Kimseyinga, Thomas Reardon, and Kyosti Pietola. "Farm Productivity in Burkina Faso: Effects of Animal Traction and Nonfarm Income." *American Journal of Agricultural Economics* 76 (1994): 608–12.

Sbacchi, Alberto. *Ethiopia under Mussolini: Fascism and the Colonial Experience*. London: Zed Books, 1985.

Sbacchi, Alberto. *Il colonialism italiano in Etiopia 1936–1940*. Milan: U. Mursia, 1980.

Scheidegger, Dino. *Äthiopien rekonstruiert: Landschafts-, Landnutzungs- und Walddynamik im Hochland von Äthiopien zwischen 1930 und 1998 anhand von Photovergleichen*. Bern: Centre for Development and Environment, Geographisches Institut, 1999.

Schwab, Peter. *Haile Selassie I: Ethiopia's Lion of Judah*. Chicago: Nelson Hall, 1979.

Scott, James. *Seeing Like a State: How Certain Schemes to Improve the Human Condition Have Failed*. New Haven, CT: Yale University Press, 1998.

Sebsebe Demissew. "A Study of the Vegetation and Floristic Composition of Southern Wallo, Ethiopia." In *African Savannas: Global Narratives and Local Knowledge of Environmental Change*, edited by Thomas J. Bassett and Donald Crummey, 159–82. Oxford: James Currey, 2003.

Seleshi Sisaye. "Urban Migration and the Labor Movement in Ethiopia." In *Proceedings of the Fifth International Conference on Ethiopian Studies: Session B April 15–16, 1978, Chicago, USA*, edited by Robert Hess, 679–700. Chicago: University of Illinois at Chicago Circle, 1979.

Sen, Amartya. *Poverty and Famines: An Essay on Entitlement and Deprivation*. Oxford: Clarendon Press, 1981.

Shepherd, Jack. *The Politics of Starvation*. New York: Carnegie Endowment for International Peace, 1975.

Shiferaw Bekele. *An Economic History of Modern Ethiopia*. Vol. 1, *The Imperial Era, 1941–74*. Dakar: CODESRIA, 1995.

Simoons, Frederick J. *Northwest Ethiopia: Peoples and Economy*. Madison: University of Wisconsin Press, 1960.

Spinage, C. A. *Cattle Plague: A History*. New York: Kluwer Academic/Plenum, 2003.

Stahl, Michael. "Environmental Degradation and Political Constraints in Ethiopia." In *Ecology and Politics: Environmental Stress and Security*, edited by Hjort af Ornäs and M. A. Mohamed Salih, 181–96. Uppsala: Scandinavian Institute of African Studies, 1989.

———. "Environmental Degradation and Political Constraints in Ethiopia." *Disasters* 14, no. 2 (1990): 140–50.

Stähli, P. "Changes in Settlement and Land-Use in Simen, Ethiopia, Especially from 1954 to 1976." In *Simien Mountains: Ethiopia*, vol. 1, *Cartography and Its Application for Geographical and Ecological Problems*, edited by B. Messerli and K. Aerni, 33–70. Bern: Geographishches Institut der Universität Bern, 1978.

Stähli, P., and M. Zurbuchen. "Two Topographic Maps 1:25,000 of Simien, Ethiopia." In *Simien Mountains: Ethiopia*, vol. 1, *Cartography and Its Application for Geographical and Ecological Problems*, ed. B. Messerli and K. Aerni, 11–31. Bern: Geographishches Institut der Universität Bern, 1978.

Sutcliffe, J. P. "Soil Conservation and Land Tenure in Highland Ethiopia." *Ethiopian Journal of Development Research* 17, no. 1 (April 1995): 63–88.

Tadesse, M., B. Alemu, G. Bekele, T. Tebikew, J. Chamberlin, and T. Benson. *Atlas of the Ethiopian Rural Economy*. Washington, DC: International Food Policy Research Institute; Addis Ababa, Ethiopia: Ethiopia, Central Statistical Agency, 2006.

Teferi Abate. "Land Scarcity and Landlessness in North Shewa: A Case Study from Wayu and Anget Mewgiya PA." In *Land Tenure and Land Policy in Ethiopia after the Derg: Proceedings of the Second Workshop of the Land Tenure Project*, edited by Dessalegn Rahmato, 95–132. Trondheim: University of Trondheim, Centre for Environment and Development, 1994.

Tesfaye Wolde Medhin. "Highland Farmers and the 'Modernizing' State in Ethiopia: Conjunctures and Disjunctures." PhD diss., University of Illinois at Urbana-Champaign, 2004.

Tiffen, Mary, Michael Mortimore, and Francis Gichuki. *More People, Less Erosion: Environmental Recovery in Kenya*. London: John Wiley, 1994.

Traversi, Leopoldo. *Let-Marefià: Prima Stazione geografica Italiana nelo Scioa e le nostre Relazioni con l'Etiopia (1876–1896)*. Milan, 1931.

Traversi, Leopoldo. Lat Marafeya, January 14, 1892. *Bolletino della Società Geografica Italiana*, series 3, vol. 5 (1892): 228–30.

Tucket, C., and H. D. Newcomb. "Expansion and Contraction of the Sahara Desert from 1980 to 1990." *Science* 253 (1991): 299–301.

United Nations Environment Programme. *Report of the United Nations Conference on Desertification, 29 August–9 September 1977*. Nairobi: United Nations Environment Programme, 1977.

Vignéras, Sylvain. *Une Mission française en Abyssinie*. Paris, 1897.

Voellmy, J., ed. *Arthur Rimbaud: Correspondance 1888–1891*. Paris: Gallimard, 1965.

von Braun, J. *A Policy Agenda for Famine Prevention in Africa*. Food Policy Statement no. 13. Washington, DC: International Food Policy Research Institute, 1991.

von Breitenbach, F. "National Forestry Development Planning: A Feasibility and Priority Study on the Example of Ethiopia." *Ethiopian Forestry Review* 3 (1962): 41–68.

Waller, Richard. "Emutai: Crisis and Response in Maasailand 1883–1902." In *Ecology of Survival: Case Studies from Northeast African History*, edited by Douglas Johnson and David Anderson, 73–112. London: I.B. Tauris, 1991.

Webb, Patrick, and Joachim von Braun. *Famine and Food Security in Ethiopia: Lessons for Africa*. Chichester: John Wiley & Sons, 1994.

Westphal, E., and J. M. C. Westphal-Stevels. *Agricultural Systems in Ethiopia.* Wageningen: Centre for Agricultural Publishing and Documentation, 1975.

Williams, Timothy O. "Problems and Prospects in the Utilization of Animal Traction in Semi-Arid West Africa: Evidence from Niger." *Soil and Tillage Research* 42, no. 4 (1997): 295–311.

Winter-Nelson, Alex. "Rural Taxation in Ethiopia, 1981–1989: A Policy Analysis Matrix Assessment for Net Producers and Net Consumers." *Food Policy* 22 (1997): 419–32.

Winqvist, E. *Med livet som insats. Läkaremissionen i Eritrea.* Stockholm, 1945.

Winqvist, E. "En liten återblick på min 25-åriga verkshamet som läkaremissionär." *Varde Ljus* (1912): 38–61.

Wiseberg, L. "An International Perspective on the African Famines." In *The Politics of Natural Disaster: The Case of the Sahel Drought,* edited by M. H. Glantz, 101–27. New York: Praeger, 1976.

Wøien, Halvor. *Woody Plant Cover and Farming Compound Distribution on the Mafud Escarpment, Ethiopia: An Aerial Photo Interpretation of Changes 1957–1986.* Working Paper on Ethiopian Development no. 9. Trondheim: University of Trondheim, Centre for Environment and Development, 1995.

Wolde Michael Kelecha. *A Glossary of Ethiopian Plant Names.* 4th ed. Addis Ababa: Addis Ababa University Press, 1987.

Woldeamlak Bewket. "Household Level Tree Planting and Its Implications for Environmental Management in the Northwestern Highlands of Ethiopia: A Case Study in the Chemoga Watershed, Blue Nile Basin." *Land Degradation and Development* 14 (2003): 377–88.

Woldeamlak Bewket and Declan Conway. "A Note on the Temporal and Spatial Variability of Rainfall in the Drought-Prone Amhara Region of Ethiopia." *International Journal of Climatology* 27 (2007): 1467–77.

Workineh Degefu. "Some Aspects of Meteorological Drought in Ethiopia." In *Drought and Famine in Africa: Denying Famine a Future,* edited by Michael Glantz, 23–26. Cambridge: Cambridge University Press, 1987.

Worster, Donald. *Dust Bowl: The Southern Plains in the 1930s.* Oxford: Oxford University Press, 1979.

Wright, Marcia. "East Africa, 1870–1905." In *Cambridge History of Africa,* vol. 6, *From 1870 to 1905,* edited by Roland Oliver and G. N. Sanderson, 539–91. Cambridge: Cambridge University Press, 1985.

Wudu Kassu. "References to the Environment in Nineteenth Century Ethiopian Chronicles." Unpublished paper.

Wurtz, R. "Hygiène Publique et Privée en Abyssinie." *La Semaine Médicale* (1898): 489–94.

Wylde, Augustus B. *Modern Abyssinia.* London, 1901.

Yeraswork Admassie. *Twenty Years to Nowhere: Property Rights, Land Management and Conservation in Ethiopia.* Lawrenceville, NJ: Red Sea Press, 2000.

Yeraswork Admassie, Mulugetta Abebe, Markos Ezra, and J. Gay. "Report on the Sociological Survey and Sociological Considerations in Preparing a Development Strategy." Institute of Development Research, Addis Ababa University, December 1983.

Yilma Seleshi and Ulrich Zanke. "Recent Changes in Rainfall and Rainy Days in Ethiopia." *International Journal of Climatology* 24 (2004): 973–83.

Young, John. *Peasant Revolution in Ethiopia: The Tigray People's Liberation Front, 1975–1991*. Cambridge: Cambridge University Press, 1997.

Zergaw Asfera. "Some Aspects of Historical Development in 'Amhara, Wallo,' *ca.* 1700–1815." Bachelor's thesis, Addis Ababa University, 1973.

Zewde Tessema Segele and Peter J. Lamb. "Interannual Variability of Growing Season over Drought-Prone Areas of Ethiopia." *CLIVAR Exchanges* 8, nos. 2/3 (2003): 36–39.

INDEX

Page numbers in italics refer to figures and tables.

Africa and the Diaspora: History, Politics, Culture

Thomas Spear, Neil Kodesh, Tejumola Olaniyan, Michael G. Schatzberg, and James H. Sweet

Series Editors

Postcolonial Paris: Fictions of Intimacy in the City of Light
LAILA AMINE

Spirit, Structure, and Flesh: Gendered Experiences in African Instituted Churches among the Yoruba of Nigeria
DEIDRE HELEN CRUMBLEY

Farming and Famine: Landscape Vulnerability in Northeast Ethiopia
DONALD CRUMMEY; edited by JAMES C. MCCANN

A Hill among a Thousand: Transformations and Ruptures in Rural Rwanda
DANIELLE DE LAME

Spirit Children: Illness, Poverty, and Infanticide in Northern Ghana
AARON R. DENHAM

Defeat Is the Only Bad News: Rwanda under Musinga, 1896–1931
ALISON LIEBHAFSKY DES FORGES; edited by DAVID NEWBURY

Education as Politics: Colonial Schooling and Political Debate in Senegal, 1850s–1914
KELLY M. DUKE BRYANT

Power in Colonial Africa: Conflict and Discourse in Lesotho, 1870–1960
ELIZABETH A. ELDREDGE

Nachituti's Gift: Economy, Society, and Environment in Central Africa
DAVID M. GORDON

Intermediaries, Interpreters, and Clerks: African Employees in the Making of Colonial Africa
Edited by BENJAMIN N. LAWRANCE, EMILY LYNN OSBORN, and RICHARD L. ROBERTS

Naming Colonialism: History and Collective Memory in the Congo, 1870–1960
OSUMAKA LIKAKA

Early African Entertainments Abroad: From the Hottentot Venus to Africa's First Olympians
BERNTH LINDFORS

Mau Mau's Children: The Making of Kenya's Postcolonial Elite
DAVID P. SANDGREN

Whispering Truth to Power: Everyday Resistance to Reconciliation in Postgenocide Rwanda
SUSAN THOMSON

Antecedents to Modern Rwanda: The Nyiginya Kingdom
JAN VANSINA

Being Colonized: The Kuba Experience in Rural Congo, 1880–1960
JAN VANSINA

Kongo in the Age of Empire, 1860–1913: The Breakdown of a Moral Order
JELMER VOS

The Postcolonial State in Africa: Fifty Years of Independence, 1960–2010
CRAWFORD YOUNG